T0348812

Synthesis Lectures on Computer Science

The series publishes short books on general computer science topics that will appeal to advanced students, researchers, and practitioners in a variety of areas within computer science.

Norman I. Badler

On Raising a Digital Human

A Personal Evolution

 Springer

Norman I. Badler (iD)
Cesium GS and The University of Pennsylvania
Philadelphia, PA, USA

ISSN 1932-1228   ISSN 1932-1686 (electronic)
Synthesis Lectures on Computer Science
ISBN 978-3-031-63944-9   ISBN 978-3-031-63945-6 (eBook)
https://doi.org/10.1007/978-3-031-63945-6

This Springer imprint is published by the registered company Springer Nature Switzerland AG
The registered company address is: Gewerbestrasse 11, 6330 Cham, Switzerland

To my colleagues, students, postdocs, staff, and family who have made this story possible, exciting, and meaningful

Foreword

Modeling virtual humans is a critical capability for engineers and designers to incorporate real-time interactive manipulation and display of human figures into computer-aided design systems. For design engineers, such models can also inform on how to control and design human-centered systems central to the emergent concept of concurrent engineering. Additionally, modeling a large group of virtual humans, i.e., computational methods and models for virtual crowd simulation of agents with artificial intelligence (AI), is considered one of the essential components of reproducing and evaluating egress performance in specific scenarios to improve human safety and comfort in built environments as well as for creating believable virtual environments for numerous applications. For example, understanding pedestrian behaviors during egress situations is of considerable importance in such contexts. Evaluation, analysis, and comparisons of crowd simulation data, derived from real-world experiments, are often required for a wide range of architectural and engineering designs in large-scale structures and/or engineered artifacts (e.g., buildings, architectures, bridges, aircraft, trains, public transit systems, etc.) and also for digital production of online educational materials and story-telling.

Dr. Norman (Norm) Badler is one of the earliest Computer Graphics pioneers who led the research in Computer Animation for human modeling and crowd simulation. For nearly five decades, his research has focused on designing algorithms and developing computer software for computer graphics with a focus on *modeling virtual humans*. He is widely known for his extensive work on 3D computer graphics representations of the human body and his expansive research to acquire, simulate, animate, and control human body, face, gesture, locomotion, and manual task motions. These virtual humans are intended to portray physical, cognitive, perceptual, personality, relationship, and cultural parameters and are used singly, in small groups, or in crowd masses. These digital models exhibit physiological, self-initiated (autonomous), task-dependent, reactive, interpersonal, or social behaviors. They are presented through 3D computer graphics systems that allow interactive visual experiences suitable for training or real-time immersion in populated virtual worlds.

Beyond the science and engineering of digital humans and virtual crowds, the personal narrative of this book is a detailed chronicle of how Norm's passion for computer science and mathematics started to evolve into a thriving and distinguished career as the leading authority on virtual human modeling today. Beginning in grade school, Norm was captivated by digital computers and polyhedra, which led him to master programming by high school. He started by exploring and pursuing his interests starting at the University of California, Santa Barbara then the University of Toronto, where he finished his doctoral research before eventually establishing a distinguished academic career at the University of Pennsylvania. Growing up in an entirely modest and non-academic family, his path is a tale of luck, coincidences, circumstantial exploration, and transformative experience. Throughout, he was fortunate to encounter and collaborate with mentors, colleagues, and especially his wife, Virginia, and two sons, Jeremy and David. Along the way, Norm pioneered novel approaches, software systems, and impressive prototypes that pushed the frontier of human modeling and computer animation. He has educated generations of researchers and students, and he bridged computer graphics and other scientific and engineering disciplines, including anthropometry, cognitive science, and ergonomics. Norm's impact can be truly felt across education, entertainment, academia, and industry around the world!

By highlighting Norm's journey from elementary school to an incredibly successful career in computer science as a world-renowned leading authority in Computer Graphics, especially on modeling and animation of virtual humans, this book showcases the importance of pursuing one's passions with determination and hard work. Norm's persistence and curious mind led him to a stellar culmination of incredible achievements ranging from research to teaching, computer vision to computer graphics, digital humans to animated crowds, and articulated movement to expressive emotions and cognitive behaviors. This autobiography of Prof. Norm Badler serves as an inspiration to all of us with its many important lessons on relationships between people and across communities as well as the essential role that real, industrial-strength problems have in driving virtual human research and more.

July 2024

Ming C. Lin
Barry Mersky and Capital One E-Nnovate
Endowed Professor, Distinguished
University Professor
University of Maryland
College Park, MA, USA

Preface

This book tells the story of building digital virtual human models in the context of the background, choices, and occurrences that shaped the author's own involvement and personal evolution. Such digital models found motivating problems and applications in engineering, anthropology, medicine, and crowd simulation, and numerous connections to other disciplines have informed and enriched their design, development, and deployment.

Encouragement to tell this story originated from several sources. Dave Kasik and Mary Whitton asked me if I would like to write a personal essay sometime in 2023 on "The Story of Virtual Humans" for *IEEE Computer Graphics and Applications.* That article, for its "Origins" series, appears in the November/December 2023 issue. It is a much-abridged version of this full account, motivated by Patrick Cozzi. He independently asked if I would do a similarly themed piece as a series of blog posts in my role as Head of Research at Cesium. After I retired in 2021 from my tenured position on the Computer and Information Science faculty at the University of Pennsylvania, I had toyed with the idea of committing memories to text before they were gone forever. Personal reasons further incentivized me to start writing in late 2022. I would not have undertaken such an enterprise today had I not experienced the joy of non-technical writing for the first time when I spent summer 2017 authoring, with my wife Virginia, our book *Dachshund Days.*

I have generously used citations to provide pointers into relevant literature, fully recognize co-authors, and provide external evidence for claims. This millennium has seen an explosion of interest in virtual beings—from both the academic and corporate spheres—which would require immense effort to properly survey. This is not a survey about today, but about how we got here.

For a proper accounting, I had to reach back into grade school in the 1950s. I can then tell the full story of computers, graphics, and virtual humans within a context of events, coincidences, good fortune, and influential people who have contributed to shaping my life and career. I have received good advice and bad. Sometimes the latter, by being ignored, turned out to be the best of all. Perhaps you may even find a few useful life pointers along the way. I hope you will enjoy following my journey.

This account is for anyone who wants to understand the history of virtual digital human beings, how they evolved, and especially how they must address numerous human characteristics to achieve any sense of "human-ness." The discussions are deliberately kept accessible to a modestly technical-savvy audience, avoiding algorithmic details and limiting unnecessary jargon. Extensive references to the literature fill those gaps. It is not an "art" book, either. While the visual appearance of virtual humans can be nearly flawless today, the underlying properties and controls these figures require are the more important aspects of this book. It could be a motivator to explore the development and applications of digital humans today, whether as a user, researcher, teacher, or just as an observer of the amazing likenesses we can see in games, movies, and the Internet. Digital human background history has not yet been adequately coalesced into a single volume. This account tries to remedy that and includes a deep personal and evolutionary perspective to humanize the story.

Haverford, PA, USA   Norman I. Badler
2024

Acknowledgments

I have endeavored to recognize the numerous individuals who enabled, participated in, and fundamentally created this story. I hope I have provided fair and accurate credit and assessments and apologize to any who were accidentally omitted. By including extensive references, all contributors should be remembered for their roles. I have also tried to acknowledge, within the stories themselves, the external sponsors who provided the problems, the funding, the talk invitations, and the intellectual and engineering contexts for many of these endeavors. I have also tried to fully describe and express my appreciation for numerous University of Pennsylvania faculty collaborators over the years: Aravind Joshi, Ruzena Bajcsy, Howard Morgan, Stephen Smoliar, Bonnie Webber, Mark Steedman, Martha Palmer, Mitch Marcus, Michael Greenwald, Stephen Lane, Dimitris Metaxas, Ani Nenkova, Barry Silverman, Renata Holod, Alla Safanova, Ladislav Kavan, Chenfanfu Jiang, and Clark Erickson. I realized that I needed to add the motivational Prologue when describing the book to Penn English Prof. Al Filreis. Numerous students, staff, postdocs, and visiting colleagues all deserve thanks for the multiplicity of influences they have originated, authored, and contributed over five decades. I tried to give them their rightful credit here.

The process of writing takes time and focus. Retiring from my tenured faculty position at the University of Pennsylvania provided some time to consider writing. A huge incentive came from colleagues David Kasik and Mary Whitton, who encouraged me to document the *Jack* human modeling software story. "On Raising a Virtual Human" was subsequently published in the November/December 2023 issue of *IEEE Computer Graphics and Applications* (DOI 10.1109/MCG.2023.3320319; 0272-1716 © 2023 IEEE). This material is reprinted here with permission from IEEE.

Another great impetus came from Patrick Cozzi, CEO of the Philadelphia geospatial software company Cesium. After retiring from Penn, Cozzi hired me into a new role as Cesium Head of Research and suggested that I write a blog about my earlier work on digital humans. That idea evolved rapidly into the full story documented here. Cozzi's support and encouragement were invaluable and much appreciated.

Oddly enough, however, the commitment to write actually came with our adoption of two dachshund puppies, Oscar and Vienna, in late 2022. At the end of 2022 and into January 2023, we had as a houseguest a relative who had to be separated from them. So the dogs and I spent hours sequestered in the family room. I had dedicated time to begin writing.

Throughout my career, and as recounted here, my wife Ginny and our two sons Jeremy and David have been inspirational. I have been especially happy to recount how Ginny was not just supportive but fundamentally motivational and essential to my intellectual evolution. I am convinced my career path would have been quite different—and likely not nearly as satisfying and rewarding—had she not been by my side since our marriage in 1968. Although my parents passed away before this work was completed, I believe I inherited my algorithmic mindset from my father Bernard, and any artistic appreciation from my mother Lillian.

Special thanks are due to those who read, commented on, and corrected portions of earlier drafts. Dave Kasik, Mary Whitton, Bonnie Mitchell, and other anonymous reviewers improved my *IEEE CG&A* Origins versions, and these modifications helped here. I especially note editorial and content contributions from Andie Tursi of Cesium, Amy Calhoun of the University of Pennsylvania, Funda Durupinar of the University of Massachusetts, Bonnie Webber and Mark Steedman of the University of Edinburgh, Aline Normoyle of Bryn Mawr College, my wife Ginny, and my sons Jeremy and David Badler. I appreciate permission to use images from Self Magazine, Lynn Oseguera, and Michael Zyda, and of real people Virginia Badler, Jeffry Nimeroff, and Wang Gu.

In my quest to find a publisher, my long-time friend and prior publisher Mike Morgan proved invaluable by recommending this work to Christine Kiilerich of Springer Nature. I am grateful for her support and the Springer Nature staff in bringing this to fruition.

Contents

Prologue

"Who is real?" The question sounds odd, even vaguely ungrammatical. Using "who" implies a reference to a person; anything else fares more sensibly with "what." As we approach the quarter century mark, the question becomes meaningful and even necessary. Just because someone looks real does not necessarily imply that they are. We watch motion picture actors who are replaced by digital doubles, who may no longer be alive, or who are imaginative versions of someone who is portrayed as if they have been alive. We become immersed in the story. The acting and crafting of such re-animations are good enough to visually convince the viewer that they present as "who" and not "what." Already the technologies and skills are accessible enough for the creation of virtual news announcers, guides, weather reporters, and customer service representatives. Marketing is a key sector, as virtual influencers are always available to make a pitch and real money can flow for goods, fashions, and services. Only a few decades ago we experienced the first telephone call centers with synthetic voice-response systems for reservations and straightforward questions. We could detect almost immediately that these interlocutors were not real and that we could often escape to a live voice by pressing "0." Visual depictions of interacting virtual people now confound that simple formula. They seem to act, or at least look, like people, so our relationships with them, whether positive or negative, are influenced by visual, cognitive, and emotional cues honed over millennia of our own survival and evolution.

"Who is real?" begs the broader question of "What makes something real?" The question, and possible answers, have engaged philosophers since philosophy itself existed. Does Descartes' "I think, therefore I am" apply to a virtual *digital* being? If it has the appearance and cognitive abilities of a human, does that make it human? Current discussions debate the role and ethics of using machine learning to construct artificial intelligence systems that can compete with human intelligence. The Turing test, involving human-like conversational responses, is no longer a measure of a human intellectual presence. Nor is tangibility essential for "Who is real?" Humanoid robots are coming. They are physical and hence certainly tangible, but technological difficulties still simplify the problem of assessing reality. Something tells us they are not human. The roboticist Masahiro Mori observed that effect in 1970 with his invention of the phrase "the uncanny valley" [1]. As our artifacts come to resemble humans, they reach a point where the forced realism itself is subliminally detectable and visually disturbing. The best way to avoid the uncanny valley is to emphasize some nonhuman attributes of our reproductions. We usually have no "reality" issues with cartoons, monsters, or industrial robots. We can still attribute human qualities to their behaviors without them taking on accurate human forms.

Defining "human-ness" is way beyond my horizon. As a computer scientist, I try to reduce complexities to structures and algorithms. To start, one must decide which human dimensions are most important, contributory, or even relevant. Analysis by narrative is a rich and endless literary route for describing human characters, but it does not provide any essential insights into how to frame "realness" algorithmically. Analysis by decomposition is a possible strategy: by developing models for the various attributes and abilities that make us human, we might be able to design and build computational components that contribute to a more realistic whole. There are no guarantees that even if we succeed in modeling some dimensions, we will be able to reassemble them into a coherent, consistent, and complete being. Entire fields of human scientific endeavor, such as psychology, cognitive science, neuroscience, anthropology, and biomechanics, have attacked some of these component dimensions. As deep and powerful as these explorations are, they historically saw little need to marry their perspectives with those of the other communities. Cross-fertilization and interdisciplinary studies have emerged as viable research directions in the past few decades. Interconnections are challenging, but the re-integration of at least a portion of what is known will be increasingly motivated by emergent applications for digital humans.

Where to start? There are many dimensions to human beings, so the first task is to decide which ones are both meaningful and computationally manageable. One possible organization appears in Fig. 1.1. In addition to the obvious *Visual Realism* aspect, I chose dimensions of *Personal Familiarity* and *Interpersonal History*. I do not think these are commonly addressed, but the three together offer a characterization of many of the present applications for digital beings. Since these categories are intended for real-time roles—as opposed to pre-recorded movie characters, for example—a real person or observer is engaged in some fashion. I will call these roles, rather interchangeably, users, players, observers, or interlocutors. Usually, context informs which term would be most appropriate. The distinctions are not particularly meaningful here.

What do the labels on my dimensions mean?

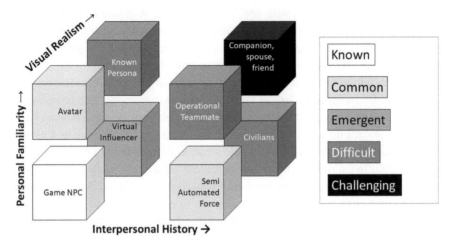

Fig. 1.1 Some virtual human dimensions and example applications

- **Visual Realism**: How important to the observer and the application are realistic human *physical* characteristics, such as shape, appearance, clothing, and faces? How important are accurate portrayals of body motions, postures, locomotion, gestures, object interactions, eye movements, and facial expressions?
- **Personal familiarity**: Does the observer *know* the virtual being? Is it a novel invention, where its background is unknown or constructed as needed? Is it a digital version of someone we've heard of or observed (perhaps through some visual media) in real life, such as an actor or sports figure? Is it someone we know quite well through first-hand experience?
- **Interpersonal history**: What *interactions* do we expect from this virtual human construction? Is the information flow one way from the virtual person to the observer, or do we share a common history, goal, or mission that requires communication, coordination, or intimacy?

In my diagram, the boxes represent selected generic applications for virtual humans. The shading indicates the approximate difficulty these applications present to the software developer. The absolute scale is mostly irrelevant; the relative position from "easy" to "hard" is important. As the applications move toward the more difficult extremes, the problems tend toward the frontiers of present software capabilities.

- A **Game NPC** is a "Non-Player-Character." These may be scripted individuals who respond in a predetermined fashion during use or game play. They may be carefully crafted and artist-designed to match the game genre. The game player may have no prior exposure to the NPC, so all information is controlled by the NPC's appearance, actions, inventory, and speech. A digital double of a motion picture actor for a film is a high-quality, artistically created, non-interactive NPC. A digital museum guide, triggered interactively by observer location, might utter pre-recorded commentary. Recent developments in artificial intelligence natural language models are beginning to impart NPCs with relevant, real-time, unscripted conversational abilities.

- **Avatars** are computational embodiments of the user. At a minimum, avatar movements are controlled through some human initiative. Avatars may appear in human form or in any other realization suited to, and reflective of, the player's persona. Avatars typically mirror user desires through a suitable interactive control interface, such as a game controller, mouse and keyboard, or wearable sensors. Camera-based systems may detect and transform facial movements and gestures into the avatar's visual representation.

- A **semi-automated force** is a collection of non-player characters, usually in a conflict situation, who behave autonomously according to engagement rules. They may have explicit higher-level goals such as attack targets or threat avoidance. Once instructed, they work on their own. They may be allies or enemies. Their behavioral rules require knowledge of the environment and other agent activities. This information may be messaged through a digital communications medium or sensed by the agent's own synthetic perceptions.

- A **virtual influencer** is an enhanced avatar, usually with an emphasis on personal appearance, contemporary notions of attractiveness, and a cultural demographic. They are usually controlled by a real person through motion capture technology, so they can act human without requiring any automated behaviors. Body shape, facial expressions, and even hair movements are important visual characteristics. Real-time interaction is often unnecessary since they can be streamed online from pre-recorded video. Influencers are often invented personalities to avoid any overt prior knowledge of their history.

- A **known persona** is a virtual version of a real, well-known individual. They may be alive at present or a digital reconstruction of a historical figure. Their history, fame, achievements, and personality are familiar enough that departures from these norms destroy the purpose and effect of virtualization. Appearance, behavior, facial expressions, gestures, clothing, and speech must all conform to their known and actual ideals.

- An **operational teammate** is an individual virtual person who works with a user to achieve common goals. Their behavioral cues might arise from player directives, but the teammate also follows any engagement rules. Some of these rules may align with their companionship obligations on a game quest. There is a growing bond between the player and teammate in their shared history. Conversational agents are a compatible term for this category when speech and gestures are the primary communicative channels.

- **Civilians** are virtual people who populate a scene. They may be passive pedestrians or other animated (even nonhuman) scene occupants, such as vehicles, ostensibly piloted by people. They should obey traffic laws and navigate complex networks of paths and roads. As virtual people, they can appear as members of a community, family, or cultural unit. The observer may feel a need to protect or avoid conflict with civilians because of these relationships. They should have enough visual realism to appear unique in dress or behavior but otherwise anonymous. The challenge lies in making these people appear to have roles, functions, and purpose within a larger environmental context.

- **Companions, spouses, and friends** are people we know very well in both public and private spheres. Virtual versions need to project mutual relationships based on intimacy, shared history, expectations, and interests. Interlocutors should expect empathy, emotional displays, voice and speech interaction, and a consistent personality. Private information must be recognized and respected during interactions. Appropriate and accurate appearance, movements, and shared interpersonal responses are crucial.

These characterizations can only address the principal attributes of these application spaces. Each has an extensive community, considerable published history, active users, observers, and collaborators. In this exposition, I cannot adequately survey the rapidly evolving state of these technologies, so I will adopt a different approach. The underlying commonality is the portrayal, in visual form, of human figures. Layered on top of that highly variable but fundamentally structured shape are its intrinsic physiological abilities, such as gestures, reaching and grasping, postural positioning, locomotion, eye movements, facial appearance, and speech. In addition, there are behaviors that organize abilities according to task achievement, communicative goals, environmental necessities, or personal maintenance. These factors, in turn, are moderated by individuality, emotion, personality, and interpersonal shared knowledge. Finally, there is the societal milieu that drives life goals, careers, ethics, action choices, and beliefs.

I have deliberately chosen to address the evolution of some of these digital human requirements as an origin story rather than a contemporary survey. I can provide considerable backstory, honor the real people who contributed to my own personal journey, and add the human experiences that often do not find their way into the technical publications that track progress. My contributions do not diminish those of many others. By the late 1980s, the virtual human animation field was already diverse, deep, and robust [2]. I merely know these stories well enough to add some life to virtual humans.

The graphic origin in Fig. 1.1 is the eponymous motivation of this narrative. Before I can address any of these dimensions or applications, I need to start with the even more basic question about why modeling humans came to be an integral part of my own life. Once aimed in that general direction, building basic human capabilities into a computer model occupied me, my colleagues, and my students for five decades. Unlike real humans, we can't start with a baby and watch it learn and grow. Moreover, unlike contemporary approaches to building computational analogs of language and image generation through machine learning and artificial intelligence methods, we did things incrementally by building models whose capabilities we understood. Our models could still exhibit interesting and novel behaviors, surprise us (and others!) with their abilities and actions, and find meaningful and rewarding roles in the service of humanity. Building a digital human ab initio, was the precursor to designing more visual realism, building better communicative partners, and making them appear to share emotions and personalities. For most of this discussion, the terms *virtual* and *digital* are interchangeable. I am not a roboticist and do not build tangible humanoid entities, but our work has influenced others who create such inventions. I arrived at a succinct summary of my professional goals in the title of a talk I gave many times in the past three decades: "My lifelong quest to control virtual people."

Origin stories are often fascinating tales of connections, discoveries, and effort. In computer graphics, we celebrate the ideas and people who bring real and imaginary worlds into the realm of synthetic visuals. Tracing the origins of new ideas by infinite regression through the citation fields of technical papers is a classic exercise for graduate students embarking on a Ph.D. Most often, however, the processes by which big ideas develop and flourish are hidden in the personal experiences of the creators and builders along the evolutionary chain. This generative process is fraught with the same transformations and consequences contributed by natural biological evolution: an unexpected mating of ideas, a meaningful mutation of a core concept, a withering of less capable offspring, and even the butterfly effect of random environmental influences. As we look backward, we can better recognize this evolutionary process to seek or at least understand the triggering conditions. While in the midst of this evolution we may not clearly see those "aha" moments that will shape the future, they may be uncovered in retrospect to trace the crucial influences along the path.

I have been extremely fortunate to have experienced and contributed to the software development of digital humans. Whether they are avatars, conversational agents, game NPCs, or virtual influencers, they all must have origins in the universe of computer graphics. Indeed, the "DNA" of these human models encompasses a large array of computer graphics component technologies, including 3D modeling, animation techniques, motion capture, and interactive systems. Finding (or purporting to know) "the" path from these origins to the present would require writing a computer graphics history book. However, what I can do and will try to document here is trace an evolutionary path—and a personal journey—to elucidate the context and processes that led to the explosion of digital human software "DNA" which has enabled contemporary embodiments for virtual beings.

References

1. M. Mori. "The Uncanny Valley: The Original Essay." June 2012. https://spectrum.ieee.org/the-uncanny-valley.
2. R. Earnshaw, N. Magnenat-Thalmann, D. Terzopoulos and D. Thalmann. "Computer Animation for Virtual Humans." IEEE Computer Graphics and Applications. 1988, pp. 20–23.

Beginnings

In 1952, when I was four years old, my parents moved to California from their family roots in Philadelphia. I grew up in the San Francisco East Bay area, a baby boomer in a young community. Neither of my parents had any postsecondary education, and neither came from any academic or technical heritage. My father was a bomber copilot in southern Europe in World War II. After the war ended, he married my mother, and they moved to Los Angeles. Soon thereafter, I was born. After a failed attempt to start a photography business, they moved back to Philadelphia and had my sister Marilyn. When I was about four years old, we moved back to California again, this time settling in San Leandro in the Bay Area. My father mostly worked as a re-upholstery salesman. Soon, he switched to being a window salesman, taking advantage of the postwar construction boom. My mother was a part-time office secretary and homemaker. She did collect books (I never saw her reading them, however), and we had a decent, if small, library of novels, war documentaries, and art books. I think I eventually read all of them.

My father liked to build things. He was an avid balsa airplane modeler, at least until he had children. With the emergence of affordable electronic components, he ordered HeathKits to assemble his own stereo system. I got to help and was quite good at soldering, recognizing components, and interpreting resistor color code bands. My father built our own stereo speaker wood enclosures in his garage workshop. (I grew up believing that garages were never used for cars, only for power tools.) His airplane model prowess had morphed into full-scale cabinet-making. My exposure to the methodical, actively planned, error-critical, and stepwise execution of these creative projects were primordial seeds for my own algorithmic thinking. Rather than airplanes, my creative hobby was HO model railroading, something that exploited similar skills.

© The Author(s), under exclusive license to Springer Nature Switzerland AG 2025
N. Badler, *On Raising a Digital Human*, Synthesis Lectures on Computer Science,
https://doi.org/10.1007/978-3-031-63945-6_2

Fortunately, in the Bay Area, we had good schools, great teachers, and even interesting neighbors. A few doors down from us lived an electrical engineer, "Jack" Jackson, who worked at Bell Telephone. One day, he brought me an issue of *Electronics Illustrated* with an article on how to build your own digital "computer" [1]. More adder than computer, it was a 6-bit binary machine with a rotary telephone dial as input. Jackson provided me with a spare telephone dial, I ordered the other parts by mail, and soon a group of sixth graders, circa 1960, was working an assembly line to wire up the six adder units. We got a few diodes inserted backward, but Jackson debugged all the circuit boards. Viola! We had a working "computer"!

Thus began the computer part of my evolution, but more technological interests would lie fallow until later in high school. While in junior high school, the "New Math" curriculum appeared as a response to the Soviet Sputnik space launch. My eighth-grade math teacher handed me New Math algebra workbooks to read and work problems on my own over the summer. That put me about a year ahead of my peers. Instead of math class in ninth grade, I became the chemistry teacher's lab assistant. I'm still surprised that I—and the school—survived that experience.

Meanwhile, my second sister, Diana, was born, and suddenly our San Leandro house was feeling cramped. My mother was working as the office manager for a friend's medical supply company, and my father was doing well enough with a construction "side gig" that we built a new house on a hillside lot in the nearby community of Castro Valley. I built a ¼" scale model of the house from the 2D plans so that my mother could better visualize her new home. Modeling seemed to be in my blood. I also shed some during the actual construction where I helped jackhammer the trenches in the hillside shale for the foundation. I also practically bathed in paint thinner every day after spending weeks staining all the exposed wood beams. We moved there starting in my 10th grade high school sophomore year.

One day, my geometry teacher pushed a form at me and told me to apply to an NSF-sponsored High School Summer Program at San Diego State College. It was a six-week in-residence program. My parents agreed. I applied, I was accepted, and in summer 1965, I went off for an enlightening and life-changing educational experience. Professor Edmund Deaton organized the topics. We had courses in formal geometry (proofs, not just constructions) and FORTRAN II programming using punched cards on an IBM 1620 computer. That course hooked me on programming, but I had no clue or guidance about what that could portend. I returned for my high school senior year. I had done well enough in math and science to be recognized with an "Engineer's Week" Award. At the ceremony, the esteemed panel asked me what I wanted to study in college: I answered, "math and computers." They all laughed at me. I went home with my fancy slide rule reward and confusion about why my interests were not being taken seriously. Maybe I had to choose one or the other. Maybe because in 1966 there were essentially no computer science programs.

For a kid who lived in the East Bay, the premier college choices were Stanford and the University of California at Berkeley. I applied to Stanford, was accepted, but had to decline, as I couldn't afford to attend. My family's finances looked good on paper, and we had just built our own custom house, but apparently the cash situation was worse than I was led to believe. This was a missed opportunity since the incipient PC community coalesced in the mid-60s at Stanford and MIT, mostly out of model railroad clubs [2]. Since that was my passion at the time, I would have been right in that formative mix. I rejected even applying to Berkeley since I did not want to be a commuting student. (We lived only about 20 minutes away.) I opted to go to UC Santa Barbara instead, mostly because we knew it was beautiful. Moreover, weren't all the UC universities equally good? For unforeseeable reasons, it turned out to be an auspicious choice in many ways.

References

1. R. Benray. "Build a Computer", Electronics Illustrated, Vol. 3(1), January 1960, pp. 65–80.
2. S. Levy. *Hackers: Heroes of the Computer Revolution*. Doubleday, New York, 1984.

UCSB and the College of Creative Studies

3

During my first quarter at UC Santa Barbara, I felt a bit lost academically. I was ostensibly a math major since I was consistently decent in those courses. I liked physics, but the approximations and assumptions were not rigid enough for my taste. Three remarkably significant events happened that first year. The first is that I met my future wife, Virginia (Ginny). Her family had moved from her childhood upbringing on Long Island, New York, to Simi Valley in southern California. Because of her intelligence, her strong education in New York, and some confounding family issues, she skipped 11th grade in high school, graduated, and then moved out of her house and into UCSB. We met a week before her 17th birthday. The second is that, shortly after arriving at UCSB, Ginny applied to a brand-new program there, the College of Creative Studies (CCS). She was accepted into the poetry major. Inspired by her initiative and her boldness, I applied to the new CCS mathematics major. As the third event, I was accepted, and we would begin our sophomore year as two new students among only 50 in a brand-new UCSB College. As a steady couple, we spent the summer between freshman and sophomore years in northern California. Ginny lived in a UC Berkeley residence dorm taking summer classes, while I stayed with my parents in Castro Valley and worked as a carpenter's aide, building a house in the Oakland hills. We were already inseparable.

We returned to UCSB for our sophomore year. Ginny stayed in the dormitory, while I shared an off-campus apartment with three other roommates. My first CCS mathematics class, run by Professor Max Weiss, was a novel experience. He talked, but he didn't lecture. We identified problems and explored topics right on the chalkboard. There were only five or six of us in that first class. If I had any pretensions of being good at math, they were completely dashed by the abilities of the other students. The Putnam Exam was an annual mathematics competition, and two of my classmates were Putnam Award winners

N. Badler, *On Raising a Digital Human*, Synthesis Lectures on Computer Science, https://doi.org/10.1007/978-3-031-63945-6_3

already: Dan Farkas and Jerry Edgar. Edgar even won it *twice*. (For comparison, I took the Putnam Exam the following year and couldn't answer a single question.). Professor Weiss had academic interests in mathematical analysis of functions (think advanced calculus), and most of what we did in his course didn't resonate with me. I did learn to think on my feet, however, probably the most useful skill of all.

CCS was created with a 1960s era vibe of open educational philosophy in which academic freedom would allow creative types to flourish. The CCS courses themselves were all graded Pass/Fail. There were few formal course requirements—two quarters in a few of the various majors within CCS. I took courses in modern art and modern architecture to satisfy my obligations in the art category. My mother's art books were a helpful preview. These courses were valuable in helping me appreciate visual aesthetics, form, and structure, especially in the 3D built environment. To satisfy my literature requirement, I took two quarters of Russian language. As I was growing up during the space race, I thought it wise to have some knowledge of the competition. In these courses, I learned that the Russian language handled verb tenses in a fundamentally different way than English. This representational distinction would prove to be a wholly unanticipated and crucial insight for my eventual Ph.D. thesis.

Among the new courses offered in CCS was one taught by David Culler. Culler was a computer network architect with a local Santa Barbara company who had arranged an early ARPANET node at UCSB. The ARPANET was the experimental precursor to today's Internet. I don't remember his class, but I do remember spending hours programming the vector character displays with their dual keyboards that were used to connect to the ARPANET at that time. I implemented and graphed in 2D all the interesting higher-order curve formulas from my favorite book, *Mathematical Models* [1].

That I knew about this book at all is a story of the unexpected. While still in grade school, I often visited my maternal grandparents who had moved out of Philadelphia and settled in an apartment not far from the San Leandro town center. During each visit, I would walk with my grandfather down the street to the public library. There, I could pick out some books to read—and I loved to read. One visit, I found this curious book with fascinating drawings and pictures of 3D polyhedra. I had never seen such forms and complexity with fundamentally understandable geometric simplicity. I soon bought my own copy. I spent a decent portion of the early 1960s building paper and cardboard models of almost all the polyhedra in the book (Fig. 3.1). I still have some of these models above my desk. My interest in graphics arose directly from this book, especially my fascination with 3D shapes.

Interests without context or motivation are wasted opportunities. Desperation is a strong impetus for novel actions. Let me elaborate. Starting in my sophomore year, my parents were experiencing financial hard times and often failed to send me rent money for my off-campus apartment. Ginny was still living and working on the UCSB campus. Besides frequently cooking for the four men sharing our apartment plus herself, she started to pay my bills. I married Ginny after my physics final exam in our sophomore

Fig. 3.1 Five cubes polyhedral model in cardboard

year. We moved into a small duplex apartment one block from the ocean. Ginny's dog from Simi joined us. I worked cleaning and flea spraying student apartments all summer to support us. Few people can match my tally of cleaning 50 neglected student kitchen ovens by hand.

At the end of my junior year, around our first anniversary, I was facing another summer. I vowed not to clean apartments again. If I couldn't find a summer job in Santa Barbara, we would have to spend the summer living with my parents in Castro Valley. My father could likely arrange another construction job. Since I was a Creative Studies math major, however, I felt that maybe there was something better to do with my education. I looked in the Santa Barbara telephone book Yellow Pages (yes, this was well before Internet search), under the most appropriate heading I could think of: "Mathematicians." There were two listings! I called them both and made appointments for interviews. The first one I visited was a "computer time-sharing" service, essentially to connect paying customers with remote computing power. They were looking for a full-time salesperson, and we agreed that this wasn't a good fit for me.

Reference

1. M. Cundy and A. P. Rollett. *Mathematical Models*, 2nd Ed., Clarendon Press, 1961.

Kramer Research and a New Trajectory   4

My second and last hope for nearby summer employment was a company called Kramer Research, located in a building called "The Riviera" in the Santa Barbara hills. A kind, early middle-age, UC Berkeley mathematics Ph.D., Henry Kramer, greeted me for my interview. He had tried teaching at Cal Polytechnic University in San Luis Obispo, California, but didn't enjoy the experience. Kramer was now running a small but funded research operation, and he was looking for a programmer! I knew FORTRAN well enough, but this project required using assembly language on an HP minicomputer. Sure, I could learn assembly! I was hired on the spot. Being able to program in 1968 was a rare skill, and as UCSB at that time had no organized computer science major at all, even local talent was sparse. Making things worse, in the 1960s, computers had negative political connotations by being associated with "The Man," corporate establishments, and faceless personal control. The mainframe computer facility at UCSB belonged to the administration and often found itself targeted by demonstrations, Molotov cocktails, and sit-ins. It was a fortuitous accident that I discovered Kramer Research at all. Later, I found out that Kramer erroneously thought I was a math graduate student applying for the job posting he took out in the UCSB Career Center. He was quite amused that the Yellow Pages listing—which he received for free from having a business telephone number—was the actual lure.

Although being a paid programmer was already a dream job, even better was the opportunity to work on a hard problem: Optical Character Recognition (OCR). Our goal was to automate reading the handwritten numerals denoting the dollar amount on personal checks. Banks employed legions of typists to read and enter these amounts manually, which were then imprinted as magnetic-inked (MICR) numbers at the check bottom. We had a custom-built flying-spot scanner based on an oscilloscope tube to image the checks, one point at a time. The computer output X and Y positions to drive the scanner beam

© The Author(s), under exclusive license to Springer Nature Switzerland AG 2025
N. Badler, *On Raising a Digital Human*, Synthesis Lectures on Computer Science,
https://doi.org/10.1007/978-3-031-63945-6_4

anywhere on the check. A photocell behind the paper sensed the transmitted light to assess whether ink attenuated the beam energy. We recognized the flaws in this design too late. We should have used reflected rather than transmitted light, and we failed to allow for different paper densities and patterning.

We also had nagging doubts about the electronic delays in the scanner control and sensing circuitry. Because I had built HeathKits, assembled an oscilloscope, and knew how to use a voltmeter, I naively (or arrogantly) thought I could help with the flying-spot scanner. While trying to track down the scanner delays, I took the driver board out of the minicomputer. When I reinserted it, I managed to reverse it. I blew out the scanner power supply and probably something else. We had to send it back to the manufacturer for repairs. Kramer could have fired me on the spot, but instead, we agreed that I would never touch computer hardware again. It was a watershed moment, and it placed me clearly in the software camp.

The computational constraints on this application were themselves formidable. The HP 2115 minicomputer had only 8K 16-bit words. Input was through a teletype keyboard and a paper tape reader. I got pretty adept at editing paper tapes with a hole punch and cellophane tape. The front computer panel had a 16 toggle switch key bank and lights for manually entering binary into memory. We needed to use that if we had to reload the boot sector on a memory wipe crash. The boot memory essentially controlled the paper tape reader. The switch bank was also useful for hot debugging since we could directly update machine code memory instructions from the toggles.

The toggle switch input motivated my first computer graphics "game." The flying-spot scanner was controlled in X and Y by pairs of octal numbers, providing a 256×256 random addressable grid. For our check reading application, we wrote programs to guide the beam so it located, and then traced, the outline of a printed numeral. An oscilloscope positioned above the computer console mirrored the scanner beam so that we could watch the output of our beam control programs. One day, I wrote a simple program that allowed a user (or two) to flip toggle switches to position the beam at the corresponding XY position. Two people worked best, as one could flip the upper 8 X bits while the other flipped the lower 8 Y bits. The computer had to track down our input point with its own "killer" spot. The computer's spot was re-aimed at each cycle toward the user point we were rapidly repositioning with switch flips. To make this more interesting, the step size along the killer's trajectory increased with each cycle. The computer thus "chased" the user's spot around the screen with increasing velocity until the distance between them reached zero. This was a frantic, completely unwinnable game, yet I remember spending many lunch breaks inevitably but happily losing to the computer. This was 1969. We preceded Pong by three years. MIT's 1962 SpaceWar! is recognized as the first computer game title [1]. Perhaps ours was the first to be played entirely with toggle switches.

Working in such a small operation encouraged daily communication with Kramer. He became more than an employer, but also a mentor and friend. He was perhaps too modest to have been a university professor; he had tried that and didn't like it. In his own small company, he could take a personal interest in the work and the people. This struck home to me on one particular occasion. I had stayed up much too late one night and, being dutiful, dragged myself into work on time. Worried that I would fall asleep on the job, I took a No-Doz tablet before I left for the office. When I arrived, I got right to work. I was debugging some assembly code, which entailed changing memory location contents through the switch register. It was tedious work, and Kramer sat down by my side to watch. I was awake, but I became fascinated by the blinking panel lights and totally lost my concentration on the problem. Kramer noticed that I was spaced out, so he asked me to take a walk with him around the Riviera campus. The fresh air cleared my mind somewhat, and I was able to get back on task. Rather than disparage me (or worse), Kramer was a "mensch" to observe and resolve my problem. I had learned a valuable management strategy: kindness. And I never took No-Doz again.

When we started this project, we had no idea how to recognize and identify what numbers the flying-spot scanner traced on the check face. We spent the summer of 1969 trying to get the system usefully operational. We designed several recognition approaches. Every time we tried something new, we had to rewrite most of the assembly code to squeeze it into 8K. We were learning pattern recognition techniques by discovery. I figured out how to optimize assembly code manually. Summer turned into fall, and by then I was the lead programmer in the office. The pay was outstanding for the time. I decided to take a leave of absence from UCSB in the Fall quarter of my senior year so I could continue working.

Kramer, my programming crew of two, and I spent December 1969 on Wall Street in New York City attempting to polish enough of a working demo to satisfy the bank sponsor. Ever the gentleman, Kramer even supported Ginny's travel to New York to stay with me for the last half of our visit. Our team occupied part of an upper floor in some skyscraper for temporary office workspace. From the beginning, we seemed like an anomaly. Coming from California, we were perennially three hours later than the other bank staff, arriving close to noon and staying later than 5 pm. This itself was odd behavior, but we did avoid the busiest subway travel times. We soon acclimatized, as we found that the restaurant lunch period downtown ended by around 2 pm. That moved us up to a more "normal" East Coast schedule. I sported a beard by then, an additional oddity for Wall Street in the 1960s. The restaurant maître d' got to know us and nicknamed me "Abraham Lincoln."

Working in the skyscraper gave me my first real human factors lesson. In the Santa Barbara office, we had complete control of our working environment. Fortunately, all of us liked working in a quiet space. With the mild Santa Barbara weather, we could open the windows and listen to the birds. In the skyscraper, however, muzak played through ceiling speakers. We found this very annoying, so we asked the building supervisor if he could turn off the muzak in our space. "Oh no," he said, "If I do that then everyone will want it turned off!" As engineers, this didn't stop us from pushing up the ceiling panels and disconnecting the speakers ourselves. I later found out that in the early days of automobile assembly lines, adding music to the factory environment increased productivity. After a while the increase tailed off, so they stopped playing the music. However, without music, productivity again increased! The change mattered more than the content. Welcome to the world of human factors!

Ultimately, the check reading project was minimally successful—OK, it really didn't work—but along the way I gained a huge appreciation for image processing and pattern recognition techniques. I had never seen these topics in my math curriculum. In one of the most selfless acts I have ever been party to, Kramer called me into his office one day and said that he was firing me. He would give me a substantial severance bonus on the condition that I go back to UCSB and finish the senior year of my undergraduate degree. It was the best offer I could have imagined. I accepted. By taking Introduction to Biology by exam, I was able to finish three quarters-worth of required courses in just two, so I could graduate in May 1970. Eight years later, his gift would cover most of the downpayment on our first house.

Senior year was half done, and my future was still unsure. There was an active lottery-driven military draft for the Vietnam War. While graduate studies seemed to be the obvious next stage of life, draft considerations circumscribed my options. Going to graduate school was no longer a sufficient cause for deferment. I zeroed in on the University of Toronto in Canada as a destination. It had graduate programs in math for me and art history for Ginny. We understood that Toronto had a good academic reputation. The most interesting person in their math department was Professor H.S.M. Coxeter, the reigning expert on polyhedra at the time. I knew of Coxeter through the *Mathematical Models* book I cherished. Perhaps I would be able to work with him and combine my two interests in math and polyhedra. I met with one of my favorite UCSB math instructors, Professor John Ernst, a topologist. I had found topology to be a nice cross between algebra and three-dimensional geometry. I told Ernst that Toronto was an exciting graduate destination for me because of Coxeter. Ernst's response was a terse "Geometry isn't fashionable in math anymore." Coxeter himself was about 65 years old at that time. I drew a draft lottery number in the low forties. We were moving to Canada for graduate studies, fashionable or not.

Reference

1. https://www.jesperjuul.net/thesis/2-historyofthecomputergame.html.

The University of Toronto   **5**

The University of Toronto accepted both Ginny and I for graduate studies. Thankfully, I received a four-year fellowship from the Canadian government. We needed to enter Canada as "Landed Immigrants" so that we could both work there as well as attend the university. Having been married two years already, we had accumulated a sizeable collection of household goods and numerous, somewhat unusual, pets. In August 1970, we rented a 24-foot U-Haul van in Santa Barbara and hitched our Volkswagen "beetle" on the back. We first drove up to the East Bay area, where we donated "31 assorted turtles" to San Francisco's Steinhart Aquarium. We then tried to optimize our cross-country route to visit important sites such as Yellowstone National Park. We arrived in Buffalo, New York, and stayed there overnight. We would have to meet with Canadian Immigration in person the next morning to obtain our Landed Immigrant status. Fortunately, we just cleared the requirements bar due to my four-year graduate fellowship, Kramer's severance nest egg, and Ginny's facility in French. We crossed the border into Canada. We had the necessary veterinarian papers for our dog, but we decided it was best not to disclose to Canadian Customs our three pet California toads riding in the U-Haul cab.

I began my math courses, but though well delivered, they seemed uninspiring. In the Winter 1971 term, I was able to take a course from Professor Coxeter. It was not the polyhedra course I was hoping for; instead, he was teaching a more algebra theory-oriented course on group presentations. Nonetheless, it was an interesting topic. One of the key concepts was using group presentations—algebraic expressions—to characterize whether a group had the mathematical property of being "simple" or not. The decision was dependent on a process called "coset enumeration": an algorithm! I met with Professor Coxeter about programming this for my class project. He agreed, but what he said in addition was even more prescient: "Computers can be useful to mathematicians." This was the

© The Author(s), under exclusive license to Springer Nature Switzerland AG 2025   19
N. Badler, *On Raising a Digital Human*, Synthesis Lectures on Computer Science,
https://doi.org/10.1007/978-3-031-63945-6_5

first time I had ever heard an academic mathematician admit to a fundamental—and positive—role for computers. Notably, Coxeter had a postdoc, Suhas Phadke, working on a conjecture that a certain set of presentations generated a new simple group. I coded up the expressions, ran it through my program, and proved the conjecture false! This wasn't good news for Phadke, but it certainly saved him further wasted effort along that particular path. Coxeter, the world-renowned mathematician, had now explicitly sanctioned my joint academic interest in math and computers and even engaged me in a real research problem.

As a mathematics Master's student, I could take a course from Toronto's Computer Science (CS) department. I found one that aligned with the topics that I struggled with at Kramer Research: artificial intelligence (AI). I knew (or rediscovered) fundamental image processing methods and sought a deeper understanding of pattern recognition processes. Pieces of a possible intellectual life started to fall into place. During my first year in Toronto, I had been feeling like the locomotive in the popular children's Golden Book, *The Little Engine that Could*, trying to scale math mountains. I kept telling myself "I think I can, I think I can,...," even though the uphill climb was getting steeper and the top of the hill was shrouded in a dense fog of what to do with a math degree. Even before I left UCSB, CCS math Professor Weiss knew me better than I realized when he summed up my situation quite well. He distinctly told me, "I've never met someone with so little ambition." Weiss had been right all along; I wasn't listening to my own inner voice.

I remember the moment of epiphany when that uphill climb ended and I turned off to ascend a new hill with nicer scenery, newer tracks, easier slopes, and a clearer, more desirable professorial summit. I had to take a required math seminar class where we each presented a research paper. I chose an algebra paper published in 1921. Although I managed to understand it, what hit me like a bombshell was that I was struggling to internalize math from 1921, and I had 50 more years of math to go! I talked with my AI professor, and I said that I could imagine working on my Ph.D. in that domain. With Ginny's insight, foresight, and encouragement, I transferred out of the mathematics department after earning my Master's degree.

I became a computer science Ph.D. student in summer 1971 with my AI course instructor, Professor John Mylopoulos, as my advisor. As an extra bonus, I avoided the required mathematics Ph.D. preliminary exams. The Computer Science department had no topical Ph.D. exams and only checked basic competency in foreign language paper translation. I managed to do well enough understanding an operations research paper written in French. I also avoided having to do the required Master's thesis in computer science by completing my math Master's, which only required coursework. Nice sidestepping. Good all around!

I had used one year of my four-year graduate fellowship for my mathematics Master's degree. I had to finish my Ph.D. within the next 3 years, including all the computer science Ph.D. course requirements. Toronto did have an undergraduate CS degree by then, so I compressed all the necessary courses into the next two years, including theory,

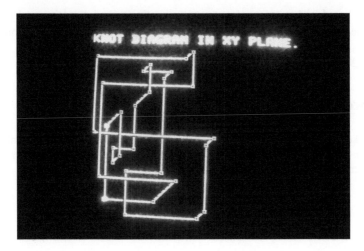

Fig. 5.1 A 3D knot diagram from a later realization of my original University of Toronto program

architecture, and information structures. All were well taught by highly esteemed professors. Perhaps the most curious of these was the information structures course taught by Canada's "Mr. Computer": Professor Calvin Gottlieb. Gottlieb's teaching style, whether deliberate or not, was to make as many mistakes as possible while lecturing and writing on the chalkboard. Because remaining alert to discover his next error captured my full attention, that was one class I could never sleep in. I didn't have the fortitude to adopt that into my own lecture style.

One of the courses I took early on was, of course, computer graphics, from famous early digital artist Professor Les Mezei. I had taken a seminar in knot theory in my last term as a math Master's student, so I did a project for Mezei's graphics course that involved displaying and unwrapping knots in 3D space (Fig. 5.1). I used an IBM mainframe computer that had an interactive vector display peripheral. I was having tremendous fun programming again.

Although Mezei was working in 2D digital media, he introduced me to key-frame animation. His work with the National Film Board of Canada animating line drawing transformations was inspiring. Computers could make *interesting* movies! In 1973, Toronto hired a new CS faculty member who had recently completed his Ph.D. in computer animation at MIT, Professor Ron Baecker [1]. I teamed up with another CS Ph.D. student, George Olshevsky, to be the Teaching Assistants for Baecker's first term computer graphics course. Olshevsky was programming visualizations of 4-dimensional polyhedra (called polytopes) for Professor Coxeter, so we were fast friends. We still have three of his four computer-drawn polytope line drawings in a six-foot-wide picture frame that hung above our fireplace mantel for many years. We could afford the framing cost, as Ginny was working for a private art gallery during our Toronto stay. Baecker and I also became friends. He broadened my view of computer animation as an interactive, real-time activity. I really liked his perspective, but I was centered on the AI representation domain, which was still young and largely open for exploration.

Reference

1. R. Baecker. "Picture-driven animation." Proc. 1969 Spring Joint Computer Conference, May 1969, pp. 273–288. (Reprinted in H. Freeman (Ed.), Tutorial and Selected Readings in Interactive Computer Graphics, IEEE Computer Society, 1980, pp. 332–347)

Component Parts for the Ph.D.   6

By 1973, I was ready to start on a Ph.D. topic. My advisor, Professor Mylopoulos, was working in artificial intelligence knowledge representation, and I started programming some of his ideas. I became facile with the SNOBOL programming language and its interpreted interactive dialect, SPITBOL (Think of these as precursors to Python, with a powerful built-in pattern-matching syntax and branch-oriented control flow.). That knowledge representation system led to my first published paper, but it was a research dead end [1]. I wanted to combine my image processing, pattern recognition, knowledge representation, computer graphics, and programming interests into something viable as a thesis. I had assembled most of the "amino acid" building blocks but had no concrete notion of how to twist them into a thesis.

Fortunately, the early 1970s presented a much more finite publication landscape than one has today. The bulk of the notable works in my interest domains came from a few top research universities, such as MIT and Stanford. I was an avid reader of all the current work out of these labs and felt increasingly certain that there was a topic to be found in computer scene analysis. At that time, considerable efforts were underway in 2D image understanding. Having already struggled with 2D OCR, I didn't find these problems particularly compelling. The work in 3D scene understanding primarily focused on recognizing 3D polyhedral shapes since line input was more copacetic to straightforward pipelines. The problems here were more inspiring to me, since 3D recognition really meant polyhedral shape reconstruction: something that computer graphics was already good at and something that I understood from the generative perspective.

To focus my thinking, I revisited the "lab notebook" paradigm that I learned to use at Kramer Research. This meant that I started to "think" directly on paper. When I read a research paper, I wrote a summary and consequent impression in the notebook in my own

© The Author(s), under exclusive license to Springer Nature Switzerland AG 2025   23
N. Badler, *On Raising a Digital Human*, Synthesis Lectures on Computer Science,
https://doi.org/10.1007/978-3-031-63945-6_6

words. There was no "cut and paste" then—this was an active transposition of thought into written words and drawings. I would speculate and ask questions of myself in the notebook and try to find possible gaps that could lead to my own topic. I spent a lot of time outdoors during the summer of 1973 in a lounge chair writing into my notebook. Nowadays, we'd probably call this activity "keeping a blog." I found the tangibility of the paper notebook medium invaluable for review and update, something we overlook today as we digest mere screenfuls of information at a time. I could write faster than I could type, but that was moot since I had no remote computer access anyway. Folks today take electronics, mobility, interaction, and Internet search for granted. Life was different in the past. I'm not saying it was better or worse, but we managed to work it to our advantage.

Reference

1. J. Mylopoulos, N. Badler, L. Melli and N. Roussopoulos. "An introduction to 1.pak, A programming language for artificial intelligence applications." Proc. Third International Joint Conf. on Artificial Intelligence, Stanford, CA, August 1973, pp. 691–695.

The ingredients in this notebook soup of ideas gradually led to a focus on 3D visual scene analysis. I did not have access to any image acquisition hardware. Work at MIT and other top universities had shown the utility of using synthesized, idealized data to bypass the messy image analysis parts. Some of them, such as work by David Waltz at MIT, used 2D projections of 3D polyhedral models as input to reconstruct the original 3D structure [1]. Not wanting to compete directly with MIT, I surmised that trying to recognize and describe what was happening with *moving* 3D polyhedra by processing their 2D projections *over time* might be an unexplored direction.

With a general idea about what I might do, I then had to ascertain the state of two crucial conditions needed to turn it into a Ph.D. The first condition was to establish that my idea wasn't already done by psychologists who, after all, had been studying human visual perception for a long time. The second condition was that another computer scientist wasn't currently doing it. To assess the first condition, I spent at least a month poring over every psychology journal in the University of Toronto library. I read fascinating works but none gave me any computationally feasible processing framework for a visual system. I did find some remarkable papers, such as Michotte's study of inferred intentions in moving 2D shapes [2] and Gibson's perceptual affordances theory [3]. I filed summaries of these away in my notebook for later use. This was, of course, long before Internet search, so these preliminary psychology investigations took real time in a real library with real books and journals. During this period, I failed to inform my Ph.D. supervisor, Mylopoulos, about what I was doing sequestered in the library. He thought I had bailed out of doing the degree and chastised me for my lack of communication. I got the message.

I still had to establish the second condition: that no computer scientist was doing what I wanted to do. The answer was simple: I would make a road trip from Toronto to MIT and assure myself by direct meetings that they—or others that they might know about—were not onto my topic yet. This would be my first "pilgrimage" to meet someone in person for advice or insight who was in an academic leadership position. By this time, I had received a draft deferment, so I could now travel into the U.S. Ginny and I drove to MIT. I was able to meet with the head of the MIT Artificial Intelligence Lab, Professor Patrick Winston. He received me cordially, and we talked about the present state of computer scene analysis at MIT. Finally, I asked him my big question: was anyone in the AI Lab at MIT working on *moving* scene analysis? His answer is another great quote that was a life changer. "Oh no," Winston said, "We don't even know how to analyze a single image yet, so we're not interested in understanding a sequence of images." The rest of that road trip must have been fun, but all I remember is that I essentially received an implicit blessing from a demigod to work on my chosen thesis topic. Now, I had to do the actual work of converting a "big" idea into a viable computational process.

References

1. D. Waltz. "Understanding line drawings of scenes with shadows." In *The Psychology of Computer Vision,* P.H. Winston (Ed.), McGraw-Hill, 1975, pp. 19–92.
2. A. Michotte. *The perception of causality.* Translated by T.R. Miles, E. Miles. Methuen, London, 1963.
3. J. Gibson. *The Senses Considered as Perceptual Systems.* Allen and Unwin, London, 1966.

The Ph.D. Thesis

8

The fundamental premise of my Ph.D. thesis is simple to state:

> People exist in a dynamically changing 3D world and effortlessly transform visual observations into textual descriptions. How can I design a computer process that achieves the same goal?

In humans, this capability is linked with language acquisition, notions of object permanence, and visual processing. Describing what we see seems to be a completely natural and well-established skill long before we undertake more structured pedagogy. I hypothesized that this descriptive facility might be computable without too many representational layers. Taking my cue from MIT, I wanted to jumpstart the process by working from synthetic visual data rather than from actual camera images.

To describe moving objects, I needed a reliable source of temporally connected image frames: a *movie*. Therefore, my first task was to implement a 3D object animation system to produce image sequences. This part was immense fun, as I could apply my graphics background. I designed and programmed a system that fed 3D object movement parameters into my own simulation engine to time-slice the movements and transform everything on a per-frame basis. Objects consisted of solid polyhedral and 3D line models. Rendering presented a novel problem (for me): since my idealized moveable point camera was a first-class 3D object, I needed to spherically project object points about the camera onto the image plane. That meant 3D straight lines mapped to arcs on the sphere. These spherical arcs were then segmented so that they could be drawn on the 2D film plane. As objects could have solid surfaces and everything existed in 3D, I needed to write a visible line algorithm to properly render these into 2D images. The image sequences were directly output onto a 16 mm microfilm plotter. This process took considerable computation time

N. Badler, *On Raising a Digital Human*, Synthesis Lectures on Computer Science,
https://doi.org/10.1007/978-3-031-63945-6_8

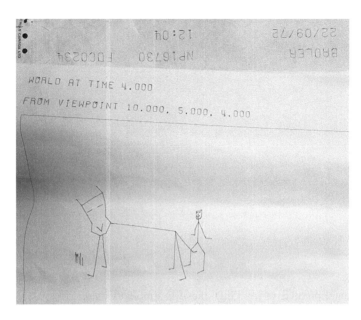

Fig. 8.1 Roll paper output from my 3D animation system. Animated cow and stick figure human. Plot date shows it was created on September 22, 1972

on the Toronto mainframe computer, and I remember that Professor Baecker came to my rescue by claiming the computation cost was for a class. In 1972, I didn't think that this animation system was itself worthy of publication, and it ended up being barely mentioned in my thesis as an appendix (Fig. 8.1). The penultimate output of the animation system (converted from film to video) is now available on YouTube [1]. The final film version of this scenario has been lost somewhere along the way to video conversion.

I gave a talk on my animation system to the Toronto CS department. I began by showing this film (Fig. 8.2), and afterward, without any prior warning, I asked everyone to write down what they just saw. That provided me with a working set of empirical descriptions with a decent amount of consistency and interesting variations. I had ground truth from the animation script. No one said they couldn't do what I asked. No one wrote more than a few terse sentences. Everyone picked up the actions. That's what verbs are for!

Now, I had to invent a verb representation that started from the 2D image coordinates and characterized object trajectories over time. Those trajectories were just ordered lists of sets of object points. But what triggered the conversion of point trajectories into verbs? The precipitating insight came from the Russian language courses I had taken in 1967 at UCSB. Russian verbs take two tense forms: *perfective* and *imperfective*. Perfective verbs describe completed actions, while imperfectives describe ongoing processes. My computational representation for an object's motion thus became a temporal list of completed

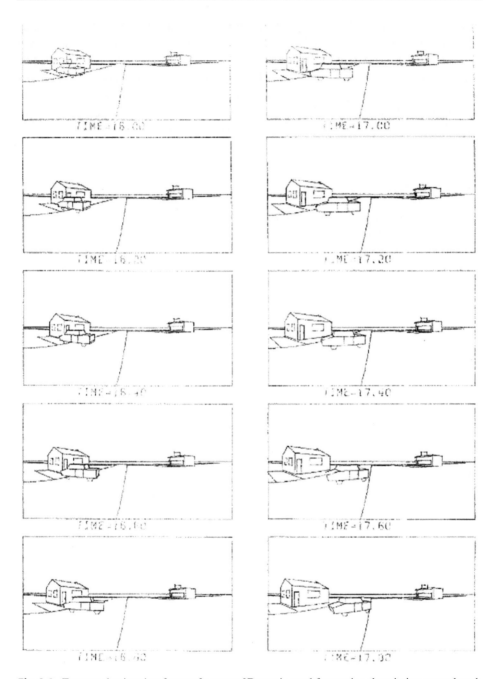

Fig. 8.2 Excerpted animation frames from my 3D movie used for motion description ground truth

actions with one possibly ongoing action. Rules for terminating a trajectory description were based on existence, continuity, and curvature conditions. A final look-up transformed the completed trajectory blocks into verbs. I wrote up the main results and submitted the paper to the Second Joint Conference on Pattern Recognition. It was accepted, and I took my first international flight to present the paper in Copenhagen, Denmark [2]. I don't remember much about the conference, but the reception at the Copenhagen City Hall was notable for the variety of hors d'oeuvres made entirely from fish.

References

1. N. Badler. https://youtu.be/UZ12RaI1Opo. Accessed March 31, 2024.
2. N. Badler. "Three-dimensional motion from two-dimensional picture sequences." Proc. Second International Joint Conf. on Pattern Recognition, Copenhagen, Denmark, August 1974, pp. 157–161.

I Need a Job, and Get an Insight, Too

9

Ginny and I truly loved living in Toronto and made many life-long friends. We rented houses and sublet to housemates. Our first house on Dovercourt Road gave us a taste of Toronto's ethnic diversity and eclecticism. The neighborhood bank branches vied with each other touting the number of foreign languages their staff spoke. Our first sublets were a colleague programmer from Kramer Research. He and his partner soon moved out, and we sublet to a couple of East Indian brothers. Soon they moved on, too. Our next housemates were a lovely Canadian couple. Martha and Richard, who would eventually be married while living with us. We experienced a Canadian wedding firsthand. The shocker for me: the beautiful white wedding cake was really a fruitcake! They, too, soon moved on to set up their own household. We then rented a house on Marmot Street where we sublet a room to one of Ginny's friends, a classics Ph.D. student, Danuta. When she moved out in 1973, we made one more hop to finally live by ourselves on Belmont Avenue. We were feeling rather "suburban" by then, enjoying a small garden and our newfound privacy. It was ideal for concentrating on the thesis.

By early 1974, I had most of the motion description pieces built but not integrated into a system. I was convinced that the parts could work together. Much of the structural representational components existed on paper rather than as code. However, a number of external factors started to press on me. My graduate fellowship had run out, and I needed to teach some courses at Toronto to earn a salary. We were expecting our first child, Jeremy, in June, so I needed a real job by the fall. I felt lulled into a sense of relative security, as I was anticipating a faculty position at Toronto. They were not averse to hiring their own Ph.D.'s at that time. I was quite disappointed to be offered only a teaching position at one of the Toronto satellite campuses. Unfortunately, or maybe out of

© The Author(s), under exclusive license to Springer Nature Switzerland AG 2025   31
N. Badler, *On Raising a Digital Human*, Synthesis Lectures on Computer Science,
https://doi.org/10.1007/978-3-031-63945-6_9

arrogance, I had not considered applying for any faculty positions in the US. The specter of desperation once more loomed on the horizon.

One day in late spring 1974, I received a letter from Professor Aravind Joshi, Chair of the Computer and Information Science (CIS) department at the University of Pennsylvania (Penn), asking me if I would like to apply for an open faculty position there. I knew very little about Penn. They were interested in me because they liked the idea of hiring a faculty member whose research fit in between Joshi's natural language processing and Professor Ruzena Bajcsy's computer vision. They only found out about me because the CS department at Toronto had sent out a list of its imminent Ph.D.'s and their abstracts to major universities. I had never even thought of applying to Penn. They had already made an offer to someone else who turned them down for an industry position. I went for an interview. Mutual desperations canceled out: I received the offer and accepted it.

Quite coincidentally, Ginny's Ph.D. supervisor in Near Eastern Archaeology was going on sabbatical in Fall 1974, and he told her to finish taking her coursework at Bryn Mawr College and the University of Pennsylvania. Penn and Bryn Mawr are only about 15 miles apart. Great! We'd move about halfway between Penn and Bryn Mawr, stay there for a year while Ginny finished her pre-thesis coursework, and then I'd get a "real" position somewhere. Well, we can see how that played out. I stayed on the Penn faculty full-time for 47 years.

I met with Joshi during my Penn faculty interview. He asked me if I was aware of recent work on motion verb semantics by Psychology Professor George Miller of Princeton University [1]. Miller's work included contributions to linguistics and cognitive science. No, I hadn't yet seen it, but promised to look at it. When I examined Miller's verb representation, I discovered the target I had been aiming for without knowing it! Miller came to a motion verb representation from the cognitive side, while I approached it from detailed geometric trajectory analysis and change conditions. The match was almost perfect. I now had a psychologically valid target for my moving scene descriptions [2]. Although I never connected any code to a natural language text generator, I was certain that such descriptions would be feasible. Moreover, this linkage validated the thesis concept I had at the very beginning: the architecture of the layers between synthetic image input and natural language output was indeed thin. Thank you, Joshi!

One interesting side effect of my verb representation was that it could be used to reason or infer motion states from a single diagram containing basic mechanical machines such as levers and pulleys. I did not publish this thesis chapter outside the document itself. That was unfortunate, as it would have predated Johan de Kleer's seminal MIT Ph.D. on qualitative reasoning by almost a year [3]. Neither of us knew about the other. There are no sour grapes here. I just became too busy with other obligations, both professional and private, to return to qualitative simulation and properly publish that work. De Kleer approached motion representation from an AI and logical reasoning perspective. Logic was not my forté, and so embedding myself into that community and its extensive literature would have been at best a diversion and at worst a mistake. Sometimes it is wiser to let things go. The qualitative reasoning community grew up without me and fostered the careers of others such as de Kleer and Ken Forbus. I would return to the qualitative simulation of fluid behaviors years later with my Ph.D. student Charles Erignac. However, it was good for me to let this direction end when it did. Recognizing one's own boundaries is a sign of maturity even if it dents one's ego.

Long after I finished my thesis, I realized that, in general, I lacked proper mentorship for a faculty research position in computer science. Writing proposals for grant support in Canada was almost perfunctory. Every faculty member seemed to have some modest research income, and all graduate students were supported through fellowships or teaching. We didn't experience proposal writing as a central mission for a faculty member. Nor were we pressed to write papers. At one point, my Toronto graph theory professor even told me point blank, "Graduate students should not be writing papers for publication; that's the faculty's job."

In my first couple of years at Penn, I spent much time sending chunks of my thesis work to lower-level, minor venues. Sometimes I was invited to do so, and sometimes I just wasn't alert to an academic publication pecking order. The graphics and AI fields were just getting started, so a lack of clear top venues was contributory. Aiming low helped improve publication success but resulted in some obscurity. One had to attend a conference and meet people to be recognized at all. In a "Hail Mary" gesture, I sent my entire thesis to a famous researcher who edited an AI book series. He rejected my thesis without substantial justification. The University of Toronto did at least publish it as a Computer Science Technical Report. They printed 300 copies, and all were soon gone. I personally mailed copies to everyone in my thesis citation list; an explicit thank you recognizing their contributions to my own.

References

1. G. Miller. "English verbs of motion: A case study in semantics and lexical memory." In *Coding Processes and Human Memory*, A. Melton and E. Martin (Eds.), V. H. Winston & Sons, Washington, D.C., 1973, pp. 335–372.
2. N. Badler. "Conceptual descriptions of physical activities." American Journal of Computational Linguistics, Microfiche 35, 1976, pp. 70–83.
3. J. de Kleer. "Qualitative and Quantitative Knowledge in Classical Mechanics." Ph.D. Thesis, MIT Computer Science, December 1, 1975.

The Penn CIS Faculty in 1974

10

We moved to Pennsylvania in late summer 1974 so I could begin teaching in the Fall term. I hadn't finished my thesis yet. One condition on my position was that I had to defend my thesis before 1975 to keep the tenure-track Assistant Professorship. If I failed to meet that deadline, the offer would revert to a non-tenure track visiting position. Fall 1974 became quite stressful between juggling a new environment, Ginny's Bryn Mawr courses, a sick baby, and finishing my thesis. Neither were these the days of "remote work" we're now all accustomed to; 1974 was pre-Internet. My thesis code and even the text existed only on a mainframe computer at the University of Toronto. I had to be physically present to use an IBM Selectric typewriter with a rotating type ball to enter and print the text. A frequent weekend activity for me that fall was an eight-hour drive to Toronto, a day of writing, typing, and editing, then another eight-hour drive back home. My deadline was closing in. I was nearly finished and started on my trip north, when my car broke down near Allentown, Pennsylvania. I had to stay in a hotel overnight while my car was being repaired. That night, in that hotel room, I wrote the final chapter of my thesis. I made the deadline, had a Ph.D. defense in mid-December 1974, formally deposited the thesis, and officially graduated in 1975. I secured my position at Penn.

CIS was one of three departments in the Moore School of Electrical Engineering at Penn. The Moore School had achieved WWII fame with the design and construction of ENIAC, an *all-electronic* digital computer. ENIAC moved out of the Moore Building in 1946. The Graduate Group in Computer Science was formed in 1959. The Graduate Group's significance was that it could legally confer the Ph.D. degree. The CIS department—as a department—began in 1972 with Professor Joshi as Chair. I honestly didn't know much about what was happening there between 1946 and 1972, other than computer

© The Author(s), under exclusive license to Springer Nature Switzerland AG 2025
N. Badler, *On Raising a Digital Human*, Synthesis Lectures on Computer Science,
https://doi.org/10.1007/978-3-031-63945-6_10

graphics Professor Andy van Dam of Brown University earned his 1965 Ph.D. degree at Penn.

When I arrived in Fall 1974, the CIS faculty was small, factional, and quirky. There were "old guard" quasi-electrical engineering Full Professors from the post-ENIAC era, a few young Assistant Professors, and some Lecturers and Research faculty. These were not like the professors I experienced at Toronto. The old guard eschewed research and even resented the very idea of it. Professor Saul Gorn, best known for his collection of self-annihilating English sentences [1], actually discouraged me from applying for more than one grant at a time, saying that I would just become a "manager." I ignored him.

A few years later, my hunch would be vindicated when I visited reigning graphics guru Professor Don Greenberg at Cornell and asked him how he built his empire. "Lots of small grants," was his immediate reply. He became my hero and implicit mentor. Professor Gorn's legacy for me was my favorite Gornism: "His research filled a much-needed gap." I vowed to be relevant in choosing research directions.

There were two other "old guard" tenured faculty, Professors John Carr and Noah Prywes, and a Research Professor, David Garfinkel. I did find out that Carr had advised a Ph.D. student who tried to use projective geometry to make 3D-like movies on an ancient DEC-338 vector display. That system was defunct by the time I came to Penn. At one time, I had a copy of that thesis and its 16 mm movie, but neither were ever published and appear lost. Most of Carr's remaining career involved being one of the first American educators to teach microprocessor technology inside China. Professor Noah Prywes spent considerable time running his company in Israel. Research Professor David Garfinkel modeled cardiac processes with differential equations. He was trying to solve or simulate them on a computer. He was more a mathematical bioengineer than a computer scientist, and he left shortly after I arrived. Lecturer David Lefkovitz taught data structures, as best I can remember, but he, too, soon left for other opportunities.

Our Chair Joshi's goal was to build up a younger group of tenure-track faculty who would refresh and lead Penn CIS into new research areas. Professor Bajcsy was his first hire: a new and early Ph.D. in computer vision from Stanford. Then, he hired Professors Robert Chen in computer architecture, Stephen Smoliar in computer music, and Peter Jessel in information systems. I was the next hire. Chen was cheerful and also useful: he was the only person who told me what I had to do to have the University know I existed and get paid. Chen left without tenure. Jessel used his Penn faculty tuition benefit to get a Wharton MBA, then quit CIS. Almost by default, my only collegial affinities were Joshi, Bajcsy, and Smoliar. Nonetheless, they each turned out to be crucially important in formulating my own research path at Penn.

The Moore School's computing environment was almost nonexistent. Although it had been the host institution for the ENIAC inventors, Penn had not capitalized on its success. There were no personal computers or workstations at that time. The main university computer was a UNIVAC machine run by a contract agency adjacent to Penn. Perhaps to the old guard, computers didn't matter. None of them seemed to program anyway. The newer faculty could program, but most had their own core teaching obligations. Thus Bajcsy had to teach the introductory programming course. Options were limited in this computing environment, and Bajcsy made a reasonable choice. She insisted that students learn *interactive* programming rather than use a punch card and teletype system. The only available interactive options were LISP and APL. LISP was not a good choice for most engineering students. Appropriate to her background in math and physics, she chose to teach APL, an IBM programming language that used a special keyboard character set. APL was an optimal choice for that time: it ran interactively from multiple terminals on the mainframe computer and was better than BASIC or FORTRAN, at least because of its powerful parallel math and matrix operators.

I immediately inherited this course from Bajcsy as my Spring 1975 semester teaching load. Although indeed interactive, APL was also a terrible choice because it was a dead end for doing any other computer science work. I soon changed the instructional language to Pascal to incorporate more diverse data structures. I spent almost two decades teaching introductory programming to hundreds of students using Pascal. I feel I paid my faculty dues by starting so many on the path of programming. An even more important personal lesson was learning how to manage a large-scale enterprise. I had numerous TAs and figured out what they needed: a "Head TA" to manage day-to-day operations. The first of these was a Ph.D. student of mine, Tamar Granor. She was a natural educator, a clear lecturer, and a great organizer. I acquired an ability to delegate authority, let competent people such as Granor do their jobs, and became a better manager. These characteristics would also prove invaluable as my research program grew and I needed someone paid to keep the enterprise on track.

Reference

1. S. Gorn. "Self-annihilating sentences: Saul Gorn's compendium of rarely used clichés." https://repository.upenn.edu/entities/publication/3758e6d1-f02b-4b65-9615-6946fbe4e412, 1992. Accessed March 31, 2024.

A Ph.D. Student!

My first year on the Penn faculty presented more challenges than just becoming a teacher. Ginny and I were juggling responsibilities with home, courses, a very young baby, and renting a house big enough for all our stuff. Although these were normal responsibilities for a young family, our son Jeremy was diagnosed with leukemia and spent months in and out of Children's Hospital at Penn. Ginny did not yet drive and was taking her required final year of Near Eastern studies coursework at both Penn and Bryn Mawr College. We seemed to be surviving, but thoughts of expanding a research program were put on a temporary hold.

Since Ginny, as an archaeologist, was a convenient collaborator, we talked about potentially interesting topics we could explore together. She mentioned that one of the sites she had been studying in class was poorly excavated, as only rough spatial position data were recorded for the found objects. Ginny wondered whether 3D computer graphics could be used to reconstruct object locations while accounting for their locational uncertainty. In 1976, during a week-long invited visit to my alma mater, UCSB, I designed and coded (in FORTRAN) the SITE system [1]. SITE allowed interactive virtual exploration of the dig site, including the extant architectural wall outlines and the database of found objects. My takeaway from this project was the empowerment that an interactive system offered to the non-computer expert. Up to this point in time, all my interactive programming experiences produced systems *for my own use*. Ginny did not need to know or learn anything about computer programming or the format of the object database. She could explore the entire site in plan or section and choose the map overlays best suited for contextual visualization. She was even able to verify an archaeological hypothesis and thereby produce a real result. We wrote a paper on the SITE system and submitted it to the annual SIGGRAPH computer graphics conference in 1978. It was accepted. For the SIGGRAPH

© The Author(s), under exclusive license to Springer Nature Switzerland AG 2025
N. Badler, *On Raising a Digital Human*, Synthesis Lectures on Computer Science,
https://doi.org/10.1007/978-3-031-63945-6_11

paper presentation, we took a picture of our 4-year-old son, Jeremy, at the console and captioned it: "So simple even a child can use it!" An early issue of *Computer Graphics World* magazine picked up the SIGGRAPH paper and reprinted it as their cover article [1]. I donated a copy to Penn's University Museum library. For a long time, if one looked for computer graphics in Archaeology, that volume appeared first.

One day, an earnest student appeared at my office door. His name was Bulent Özgüç, and he was sent over to see me by his Architecture department Ph.D. advisor, Professor Peter McCleary. Özgüç wanted to do his dissertation on 3D graphic representations for architecture, but at that time, no one in that department knew anything about computer graphics. McCleary had learned of my recent faculty appointment to Penn, possibly also heard about the SITE system and its spatial features, and suggested to Özgüç that perhaps I could be helpful. This arrangement worked fine for me, as I didn't have to fund the student and could concentrate on what he wanted to do. Özgüç received his Ph.D. in architecture in 1976—my first Ph.D. student and not even a computer scientist!

The tale wouldn't be nearly as important had it ended there with his graduation. I love the quirks and twists of accidental alignments and coincidences. Unbeknownst to me when I started working with Bulent, it turned out that his parents, Tahsin and Nimet Özgüç, were the premier Anatolian archaeologists of Turkey. Suddenly, Ginny—a nascent Ph.D. student in Near Eastern Studies—had a close connection with them through Bulent. Bulent returned to Turkey and took faculty positions in both the Computer Engineering and Architecture departments of Bilkent University in Ankara, the capital city of Turkey. A few years later, in 1993, Bulent organized the First Bilkent Conference on Computer Graphics and invited me to speak. Ginny and I went to Ankara for a delightful visit. During the Conference, a talk slot opened up when a scheduled speaker couldn't attend. Bulent invited Ginny to step in, and she gave the best talk of the conference on her discovery of the earliest chemically attested ancient wine and beer [2]!

We still had a crucial meeting ahead. Bulent, through his parents' connections, had arranged for Ginny to meet with the Turkish government's Minister of Antiquities with the purpose of securing a letter of safe passage to work on an active archaeological dig in Turkey. We met the Minister in his apartment in Ankara, had tea, and received his letter. Ginny parlayed this opportunity into three summer excavation seasons at the site of Titriş Höyük, supervised by Near Eastern archaeologist Professor Guillermo Algaze of the University of California at San Diego.

But wait, there's more! At Bilkent, Bulent began to supervise Ph.D. students of his own, including Uğur Güdükbay. After receiving his Ph.D., Professor Güdükbay would later take a sabbatical from his faculty position at Bilkent and visit me at Penn to study animated crowd models. Subsequently, Güdükbay graduated his own Ph.D. student, Funda Durupinar. In the mid-2010s, Durupinar moved to the US, and I hired her as a postdoc for about a year. By the end of that arrangement, she had produced the capstone work of my own research career [3]. I started my Penn Ph.D. family with Bulent Özgüç; I essentially closed out my virtual human research with a significant article on endowing them with motion nuances keyed by personality traits, led by my academic "great-granddaughter" Dr. Funda Durupinar.

References

1. N. Badler and V. Badler. "Interaction with a color computer graphics display system for archaeological sites," Computer Graphics 12(3), Aug. 1978. Revised, expanded version appears as the cover article in Computer Graphics World 1(11), 1978, pp. 12–18.
2. R. Michel, P. McGovern and V. Badler. "The first wine & beer." Anal. Chem. 65 (8), 1993, pp. 408A–413A.
3. F. Durupinar, M. Kapadia, S. Deutsch, M. Neff and N. Badler. "PERFORM: Perceptual Approach for Adding OCEAN Personality to Human Motion using Laban Movement Analysis." ACM Transactions on Graphics, 2017.

In 1975, I had some job security from my initial three-year faculty appointment at Penn. I had to turn my attention to establishing my own research program. Shortly after I settled in, I met with Joshi. He almost casually mentioned that surely I was aware I had a nine-month academic salary and that I wouldn't be paid in the summer unless I brought in outside research funds. So began my adventures in finding research support. Ultimately, I got pretty good at obtaining grant money. In fall 1974, I applied for and received a National Science Foundation (NSF) Research Initiation Grant based on my thesis work. Funding started in spring 1975 and covered two summer months. Yay! We would be able to eat over the summer.

With the NSF grant in hand, I was determined to restart my thesis directions at Penn. I still didn't want to descend into the image processing level, but none of my Toronto SPITBOL code would run on Penn's mainframe computer. Rethinking what I was trying to do made me realize that one of the main applications for verbal descriptions was to characterize what other *people* were doing. The animation system I created for my thesis was not really designed for animating people because it was grounded in line representations, even for solid objects.

There were good students from local industries taking Master's courses at Penn, and I soon found some helpful research candidates. Master's theses were required then—an inducement to match students with faculty interests. An amazing Master's student, Joseph O'Rourke, became one of my first Ph.D. students. Somehow I bought (or found?) a Tektronix storage tube display so we could do vector graphics off the UNIVAC mainframe. (In modern times, new faculty are traditionally given departmental "start-up" funds to get their lab off the ground. As best I can recall, when I arrived in 1974, I received a desk, a chair, three ENIAC-era wooden bookshelves, and a filing cabinet.) O'Rourke began to

© The Author(s), under exclusive license to Springer Nature Switzerland AG 2025
N. Badler, *On Raising a Digital Human*, Synthesis Lectures on Computer Science,
https://doi.org/10.1007/978-3-031-63945-6_12

code an articulated body animation system. The concept of a tree of 3D transformation matrices at joints that connect body segments was a known structure in computer graphics. In modern parlance, this construct is called a "rig" and can be designed to model all sorts of jointed and bendable characters and structures. The root node of the body rig is often a mid-torso joint, although later, we used the body center-of-mass as the root connection to the surrounding world coordinate system. Translations of the root moved the body in space. Rotations at each body joint position the rest of the parts. "Forward kinematics" is the process of computing the body pose by starting at the root and successively applying the joint transformations all the way out to the limb extremities.

Although linked structures are rather straightforward applications of nested rotation matrices, we discovered an unexpected issue. If one approaches joint motion from a robotics point of view, human naturalness in structure and ability is not essential. In the 1970s, humanoid robots were still in the future, and most robots were mechanical contrivances designed for manipulation agility and grasping functionality. A human model, however, should reflect at least some level of natural motion. One of the most interesting joints is the human shoulder complex. A simple exercise points out a difficulty with shoulder and arm animations. Stand up and place both arms straight in front with thumbs up and palms facing each other. Now, bring your left arm straight down to your side, then extend it straight out to your left, then swing it back toward your right hand. You might be a bit surprised to see that now your left palm faces down and your thumb points at your right hand. Without thinking, you have twisted your left arm 90° clockwise. Just applying joint rotations to the three orthogonal axes of the shoulder will add this unwanted twist and cause motion and positioning ambiguities. O'Rourke addressed this situation by creating a function that computed a unique default twist to every direction on the sphere about the shoulder. This defined the twist for any pose, independent of how the body arrived there. This default twist function is discontinuous along a "seam" behind the shoulder. Most people cannot reach this seam, nor can they cross it without injury.

The twist function was crucial to our development of human model reach capabilities. Reaching requires "inverse kinematics" (IK) since it proceeds opposite to the root-to-extremity positioning. Given a reach point in space for a hand, IK transforms the target's spherical coordinates to rotation matrices for the shoulder and elbow, which achieve the goal. The default twist is applied to this pose [1]. Reaches are thus consistent and repeatable wherever the reach point is positioned in space. If the reach point is outside the reachable space, the hand can be extended to its maximum and positioned on the ray from the shoulder to the target. In more advanced goal achievement models, the figure could even bend in the torso or take steps to reach the target. But I'm getting ahead of myself.

O'Rourke went on to explore how to utilize a 3D articulated body model to fit 2D projected outlines of a human image. The body segment lengths were additional fitting constraints that usually forced the 3D model to fit the 2D inputs. We achieved significant progress toward enabling a computer to observe a human's image and fit a posable

3D human model to the input [2]. As with many other Ph.D. theses, however, I allowed O'Rourke's image processing approach to lie fallow once he graduated. Subsequently, I realized that most Ph.D. students did not like to start building from the work of their predecessors, preferring to forge onward in new directions. While such attitudes are healthy and foster novelty, they eventually create code continuity and longevity issues that I could ameliorate only by hiring someone to be a lab "manager" with oversight responsibilities.

References

1. N. Badler, J. O'Rourke and B. Kaufman. "Special problems in human movement simulation." ACM SIGGRAPH Computer Graphics 14(3), Summer 1980, pp. 189–197.
2. J. O'Rourke and N. Badler. "Model-based image analysis of human motion using constraint propagation." IEEE Trans. on Pattern Analysis and Machine Intelligence 2(6), Nov. 1980, pp. 522–536.

Color Graphics! 13

Freeing ourselves from a mainframe computing platform was an important step toward research autonomy. The Penn administration's attitude toward academic computing was lax. Rather than consolidate computing resources in some large centralized facility, Penn allowed every School and department to pursue computing solutions appropriate for their local needs. Whether from mandate or neglect, this *laissez faire* policy turned out to be the best possible in the 1970s and 1980s, when consolidation at other large institutions was the norm. There would be no friction in getting our own computing platform and peripherals, as long as we could afford them. In 1977, Ruzena Bajcsy and I applied for and received an NSF Equipment Grant to purchase emergent graphics technology: a Ramtek full-color raster display device and a PDP-11 driver computer.

Bajcsy wanted to do real image processing for robot vision, and I liked the idea of doing color raster graphics after years of vector drawings. The Ramtek was good for image display but not particularly good at animation. Nonetheless, the idea of making a solid-looking human as a data source for motion description purposes was intriguing. We had no software tools available, so we had to write our own rendering code. The Ramtek did not even come with a driver to link it to the PDP-11 computer. Electrical engineering undergraduate Leonard Williams wrote one, giving us a programmable raster display system.

Rendering images of 3D polygons could be slow using this 1977-era frame buffer since the visible surfaces had to be computed on the PDP-11 and transmitted to the Ramtek. I sought a different way to exploit the raster format to obtain images that had human-like shape. Stick figures were a common visual representation of human characters because they were naturally suited to vector-drawing graphics displays. These sparse stick figures, however, obscured (that is, avoided) some hard problems. Twistable body segments, such

N. Badler, *On Raising a Digital Human*, Synthesis Lectures on Computer Science, https://doi.org/10.1007/978-3-031-63945-6_13

as the forearm, were only visible if they terminated in a hand. Often, only one waist joint is substituted for spinal flexibility. In addition, the lines oversimplified the problem of detecting self-intersections. It was relatively easy to set rotation joint limits to handle some individual joint constraints, but visualizing arbitrary segment-to-segment intersections— such as the forearm with the torso—was awkward since the segments themselves had no 3D substance.

Having a raster graphics display that allowed solid-appearing figures gave birth to our first synthetic humans: Bubbleman and Bubblewoman. Constructed with overlapping sets of spheres, the models exhibited natural solidity and a rounded body shape. This idea is derived from a well-known image processing representation called the *medial axis*: the locus of points equidistant from the boundaries of a shape. The set of sphere centers discretized and approximated the 3D medial axis transformation of a surface, such as a body segment. The tree-structured armatures on which the spheres hung were the same as those in O'Rourke's earlier model. Body segments were moved by the same 3D kinematic and inverse kinematic joint rotations. The sphere rendering was a reasonably efficient back-to-front sort of projected spheres into shaded disks with some color attenuation based on their depth from the virtual camera. More distant disks were drawn in darker shades. I even came up with an algorithm to draw 2D solid disks efficiently by minimizing data communication between the computer and the display [1]. By 1979, we had published papers on Bubblepeople with Joseph O'Rourke and Master's student Hasida Toltzis [2, 3]. We chose the *IEEE Proceedings* venue precisely because Joshi thought that a major publication there would impress the Electrical Engineering faculty in the Moore School who would be needed to support my upcoming tenure decision.

The segment twisting display problem in the forearm and torso intrigued me. Master's student Mary Ann Morris and I invented a joint rotation transformation scaled along a body segment to portray realistic shape twisting. A forearm rotation would be scaled linearly from zero twist at the elbow to the desired twist at the wrist. These scaled rotations transformed the forearm sphere centers. We could apply the same process to the torso as well. We published the algorithm in an obscure venue [4]. If only we had applied the idea to polygonal models, we would have invented *linear blend skinning*, a staple of modern shape deformation animation. (That honor is generally attributed to Nadia Magnenat-Thalmann and Daniel Thalmann [5].) Bubblewoman peaked in utility when she was featured in an exercise article in the September 1984 issue of *SELF* magazine [6] (Fig. 13.1). Master's students Marion Hamermesh and Jon Korein used Bubblewoman to replace photos of an actual model's poses, which had poor color choices and suboptimal camera angles. With our 3D graphics model, we were able to re-color and "re-shoot" the poses to better illustrate the exercises. To my knowledge, this is the first instance of a computer graphics human model starring in a major consumer-level magazine article.

These shadowy but solid-looking human figures began to attract the attention of human factors and ergonomics engineers. Through O'Rourke's work connections, we received a small U.S. Navy grant to embed Bubbleman into a 3D polyhedral workspace environment.

Fig. 13.1 The *SELF* magazine exercise article with Bubblewoman figures. Reprinted with permission

For human factors engineers, this was an enlightening application of color raster graphics to draw reasonably realistic solid renderings of how someone would fit, for example, in a cockpit (Fig. 13.2).

Around 1980, we purchased a new raster display system from Lexidata because it included the first commercially available hardware *depth buffer*. Now, we could dispense with the preliminary back-to-front sphere sort and just stream and directly render the projected disks in any order. Since changing body poses no longer required any sphere re-sorting overhead, interactivity rates improved as well. However, the added depth-buffer benefit negated the original incentive to use spheres. The Lexidata could just as readily, if not even more efficiently, consume 3D polygons to render shapes directly.

Ultimately, our choice of a sphere representation for human bodies would be superseded by the computer graphics industry-wide shift to polygonal mesh models and general visible surface algorithms. Nonetheless, spheres had some innate properties that commended them during their time. First, they conveyed a visual softness and realism in body shape, given the small number of actual spherical primitives (approximately 300), as the *Self* magazine images portray. Second, they implicitly solved a difficult shape deformation problem that plagued early polyhedral human models: twisted mesh segments, such as the forearm, could collapse into an ugly ribbon in the middle. Later work (by

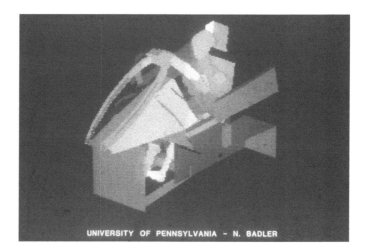

Fig. 13.2 Bubbleman in a polygonal cockpit model. SIGGRAPH'80 Program credits image and SIGGRAPH Slide set 13 (4) #72

others) on linear blend skinning would resolve that polygonal issue satisfactorily. With Morris' interpolated sphere centers along a twisted segment, this was simply not a problem. Third, the problem of self-collision across nonadjacent body segments was easy to detect as any unwanted intersection of sphere sets, themselves just a collection of distance measurements. In a polygon model, geometric interpenetration tests are complicated by the variety of vertex, edge, and face intersection combinations. Fourth, the bodies had a vaguely unclothed look; at best, they appeared to be wearing close-fitting leotards. Adding clothing and other accessories required polygonal meshes and surfaces that just weren't conducive to sphere set models. The exercise images were, perhaps accidentally, the ideal application. The benefits of using spheres were their visual novelty, the catchy name, and the community recognition that something potentially interesting was happening at the University of Pennsylvania.

The sphere model also led to a straightforward temporal anti-aliasing (motion blur) algorithm for Bubblepeople animation (Fig. 13.3). Blurred motion in an animation frame avoided a choppy or stroboscopic look. Jon Korein's process limited blurs within a frame to linear paths for the sphere centers, but there was no practical reason why curved paths couldn't be accommodated [7]. Because motion picture film cameras use a shutter with a finite opening time, they naturally contain per-frame motion blur. Cartoon animators learned to use speed line conventions to portray fast movement in a static frame. The human visual system readily accepts motion blur. A series of crisp images presented stroboscopically could be jarring and, for some people, even cause epileptic seizures. In the 1980s, however, film rates of 24 frames/second or video refresh rates of 30 frames/ second made temporal anti-aliasing a computer animation necessity. Somewhat motivated

Fig. 13.3 Bubbleman throwing a ball using temporal anti-aliasing to smooth and motion blur the individual frame images. Top, no anti-aliasing; middle, unifom filter; bottom, end pose-weighted filter

by gaming requirements, refresh rates for displays pushed toward 60 frames/second, or even higher, bypassing some of the visual incentives for temporal anti-aliasing.

After presenting Korein's paper at SIGGRAPH'83, we were congratulated by some researchers from LucasFilm (later to become Pixar). They told us they had worked on a similar problem the year before to temporal anti-alias the rapidly moving point-light starfield images required in the computer animated Genesis sequence for *Star Trek II: The Wrath of Kahn* [8]. They had not yet published their methods, but they would the following year. At the 1984 SIGGRAPH Conference, temporal anti-aliasing was a component of the seminal computer graphics process called "distributed ray tracing" [9]. Their technique added the temporal dimension to the nascent ray-tracing scheme invented by Turner Whitted just a few years earlier [10]. Our specialized approach was instantly outdated.

References

1. N. Badler. "Disk generators for a raster display device," Computer Graphics and Image Processing 6, 1977, pp. 589–593.
2. N. Badler, J. O'Rourke and H. Toltzis. "A spherical representation of a human body for visualizing movement." IEEE Proceedings 67(10), Oct. 1979, pp. 1397–1403.
3. J. O'Rourke and N. Badler. "Decomposition of three-dimensional objects into spheres." IEEE Trans. on Pattern Analysis and Machine Intelligence 1(3), July 1979, pp. 295–305.
4. N. Badler and M. Morris. "Modelling flexible articulated objects." Proc. Computer Graphics, ONLINE Conferences, Northwood, UK, 1982, pp. 305–314.
5. N. Magnenat-Thalmann, R. Laperrière and D. Thalmann. "Joint-dependent local deformations for hand animation and object grasping." Graphics Interface, 1988, pp. 26–33.
6. https://www.cis.upenn.edu/~badler/3DBubblewomanSELFMagazineSeptember1984.htm.
7. J. Korein and N. Badler. "Temporal anti-aliasing in computer generated animation." SIGGRAPH '83: Proceedings of the 10th annual conference on Computer Graphics and Interactive Techniques, July 1983, pp. 377–388.
8. Lucasfilm. https://www.ilm.com/vfx/star-trek-ii-the-wrath-of-khan/. Accessed August 27, 2023.
9. R. Cook, T. Porter and L. Carpenter. "Distributed ray tracing." ACM SIGGRAPH Computer Graphics, Vol. 18 (3), July 1984, pp 137–145.
10. T. Whitted. "An improved illumination model for shaded display." Comm. of the ACM, Vol. 23 (6), June 1980, pp. 343–349.

NASA, TEMPUS, and Beyond

14

1980 was an auspicious year. I was granted tenure and promoted to Associate Professor, and our second son, David, was born. With tenure, I could focus my research on a longer-term and more expansive view of human modeling. A new student added face animation to my interests. Steve Platt, first my Master's, then my Ph.D. student, constructed a polyhedral facial model and used physics-based springs to animate the skin mesh [1]. We were beginning to see the computational and display advantages of polygons over spheres. O'Rourke, Platt, and Morris summarized our work on human movement animation for the First Annual National Conference on Artificial Intelligence [2]. The conference was somewhat peculiar in that it had oral presentations for only some of the accepted and published papers. Our paper was print-only. During my paper session, I was sitting in the audience next to Texas A&M Professor Bruce McCormick, a computer vision researcher whose work I followed. In person, he was a substantial individual with a bear-like deportment. I turned to him and complained that I thought our paper was at least as worthy of oral presentation as the ones selected. He responded, "You have to growl a lot." I got that message clearly: if you want to be known, you can't wait for others to recognize your worth.

Perhaps coincidentally, growling or not, that paper proved pivotal in my research career. After our paper's conference session, a gentleman named Jim Lewis came up to me and introduced himself. He had read our paper, even though we didn't get to present it orally. He was from NASA Johnson Space Center (JSC) and said his Crew Station Design Section was interested in human model visualization for the Space Shuttle. I soon visited Houston JSC and formulated a plan for a software tool to aid their human factors studies. It would be the first of many enlightening and fruitful trips to Houston to work with NASA.

N. Badler, *On Raising a Digital Human*, Synthesis Lectures on Computer Science,
https://doi.org/10.1007/978-3-031-63945-6_14

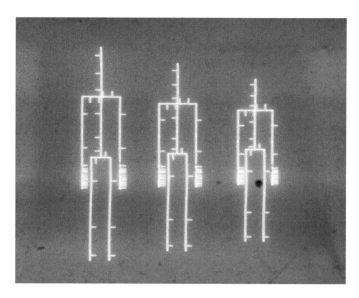

Fig. 14.1 Three anthropometrically scaled skeletons for TEMPUS: 95th, 50th, and 5th percentile males. The lines represent segment lengths, while the short "stubs" are the segment normals used as axes by the joint rotation matrices. The joint locations are indicated by the barely visible lighter dots. Points at the extremities are used for inverse kinematics positioning

In 1981, we received our first grant from NASA JSC. We agreed to provide an *interactive color* software system to evaluate human factors questions such as fit, reachability, and clearance. Among the features we proposed to build into this new software system, called TEMPUS, were polygonal human models, anthropometric scalability (Fig. 14.1), inverse kinematics for limb positioning, color raster graphics display, and real-time interaction in a 3D polygonal environment.

Having a grant from NASA to build software meant that we needed a computer. The Crew Station Design Section had a Digital Equipment Corporation VAX computer. They used it to create 3D models of the Shuttle components and drew 3D hidden-line-removed images on a plotter for hardcopies. Visible line algorithms are slow, and the Shuttle complexity meant that images often had to run overnight to produce drawings by the morning. Our own PDP-11 Ramtek driver had insufficient power or space for any real application systems. Using the UNIVAC mainframe accessible to us in the Engineering School was inadequate for the necessary interactive 3D graphics required for TEMPUS. We had insufficient funds to purchase a VAX compatible with the NASA system. We were stymied.

The VAX was a favorite research platform and an effective precursor to personal computers. Joshi had used his own grant money to purchase a VAX computer for natural language research. Joshi's VAX lived next door to the graphics lab in the computer science building basement. I asked Joshi if we could use some time on his VAX to develop TEM-PUS. Consistent with Joshi's generous nature, he said yes. But he may have had unspoken regrets, as from then on I don't know how much time he got on his own computer. We mostly took it over for graphics. Since the VAX supported real interface boards, we ordered a new color graphics display from Grinnell with 512×512 resolution and 24-bit color.

To construct the polygon models, we needed a 3D modeling system that would support joints and object hierarchies. There was no off-the-shelf modeling system in the lab in 1980, so we had to build our own. (We knew about the MovieBYU graphics software [3], but it did not have the requisite joint rotation animation structures needed to support animated characters.) Ph.D. student James Korein and Master's student Graham Walters jointly authored their own 3D modeling representation called "Peabody." Objects were first drawn on paper, and 3D coordinates were hand-entered into text files. Custom data converters eventually loaded existing formatted computer-aided design (CAD) files. Walters moved to Pixar after graduation and eventually became a Producer for the movie "Finding Nemo."

Master's student Jane Rovins constructed our first polygonal human model suitable for the raster display (Fig. 14.2). Built in Peabody and composed of triangles, the androgynous tetrahedral segment figure used the minimum shapes necessary to show solid forms in the torso, head, and limbs. It had articulated joints centered on the pyramidal apexes. It bears a remarkable resemblance to the contemporary wireframe skeleton rigs used in 3D modeling tools such as Autodesk Maya (This may be just a coincidence, but Maya precursor Wavefront Technologies did know about our software during its rise in the 1980s.). Art student Carolyn Brown created a more solid-looking "shell" figure without stressing

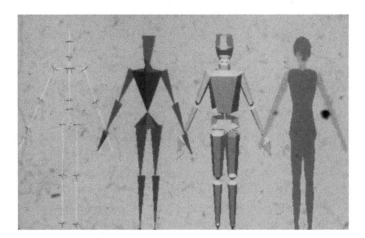

Fig. 14.2 Four versions of the human model (left to right): Articulated skeleton, triangle polygon, shell body, Bubblewoman

Fig. 14.3 A rather whimsical line-up of anthropometrically scaled male figures with polygon shells

our polygon budget. Other better-shaped models soon followed as processing limitations eased with newer graphics hardware systems.

Within a year of starting work with NASA, we delivered mostly working software. The Korein brothers, undergraduate computer science student Marie D'Amico, and I spent several days in Houston installing, debugging, and training the JSC staff to use it. We met all our promises over the following year. TEMPUS was an interactive tool engineers could now use to check fit, reach, and clearances in the space shuttle cockpit and payload bay with both "shirt-sleeved" and EVA-suited articulated human figures [4]. We were told that the TEMPUS software passed its first real use case. Congress had mandated that the Shuttle be designed for a maximally diverse crew, so it had to accommodate a full range of men's and women's body sizes from the 5th to the 95th percentile (Fig. 14.3). On early shuttle flights, there were apparently toilet issues for large male crewmembers. Using TEMPUS, the toilet configuration was redesigned and verified to better support the required diversity. This now gave me a great line for invited talks: "At the University of Pennsylvania computer graphics lab, we start at the bottom."

As part of the NASA TEMPUS project funding, we were obliged to provide the program managers with written updates. Thus began a long series of Penn Graphics Research Lab *Quarterly Reports*. These soon became the de facto reporting vehicle for all our lab projects, and we mailed bound hardcopies to all our sponsors and other interested parties. These kept everyone equally informed of our projects, provided the necessary documentation, cross-fertilized existing supporters, and seeded new projects. Since I assembled and edited each issue, I kept myself abreast of our own progress. We produced *Quarterly Reports* for about ten years, throughout the 1980s.

The TEMPUS system, our papers, and our successes began to attract further attention in the human factors arena. I had already met many seminal human model developers and readily convinced them to contribute papers for a November 1982 Special Issue of *IEEE Computer Graphics and Applications*. Among these contributions was a paper by William Fetter of Boeing, who popularized the term "computer graphics." Fetter had created an early line representation and animations of a movable body for human factors analyses [5]. The history Fetter documented provides an essential early strand of digital human DNA. There were a number of corporate human modeling efforts by then, almost all in the vehicle and cockpit design space, implemented on expensive vector-drawing displays and proprietary CAD systems.

In the early 1980s, the computer graphics community, while growing steadily, was still small. One day in 1982, I received a phone call from Gary Demos asking if I would like to help with some anticipated simulations they were undertaking at a new company, Digital Productions (DP). Oh, and did I have any knowledgeable students they could hire? I recommended three Master's students: Mary Ann Morris (who had worked on the Bubblepeople), Andy Davidson, and Emily Nagle. Demos hired all three, and I started as a consultant.

DP began graphics special effects work on *The Last Starfighter* movie. The script originally called for some computer-animated aliens. As the movie went into production, they abandoned the primary scene justifying these synthesized creatures. It was still instructive to have an inside glimpse of a nascent computer graphics effects company staffed with highly talented people. In 1983, after I made several visits to their Los Angeles facility, Demos asked me if I wanted to join DP full time. Ginny and I did not want to move to Los Angeles, especially as we had two young children. My career was going well at Penn with the TEMPUS project and new graduate students. Of course, "remote work" was not yet an option. Consulting at DP did help me realize that my future would not be as a programmer, but rather more as a research manager. With a slight tinge of regret, I turned down Demos' verbal offer. By 1987, Digital Productions folded and the staff dispersed. I was sorry to hear of the company's demise, but felt glad I had not uprooted my family.

References

1. S. Platt and N. Badler. "Animating facial expressions." ACM SIGGRAPH Computer Graphics 15(3), Aug. 1981, pp. 245–252.
2. N. Badler, J. O'Rourke, S. Platt and M. Morris. "Human movement understanding: A variety of perspectives." Proc. First Annual National Conf. on Artificial Intelligence, Stanford, CA, August 1980, pp. 53–55.
3. H. Christiansen and M. Stephenson. "Overview of the MOVIE.BYU Software System." SAE Transactions, Vol. 93, Section 3: 840403—840770 (1984), pp. 1044–1057.

4. N. Badler, B. Webber, J. Korein and J. Korein. "TEMPUS: A system for the design and simulation of mobile agents in a workstation and task environment." Proc. Trends and Applications Conference, Washington, DC, 1983, pp. 263–269.
5. W. Fetter, "A progression of human figures simulated by computer graphics." IEEE Computer Graphics and Applications, Vol. 2(09), 1982, pp. 9–13.

Building the Computer Graphics Research Lab

15

During the early TEMPUS years, the students tasked with writing code shared a small office in the Graduate wing of the Moore School. We managed to squeeze in four or five students, their desks, and computer terminals. They had to climb over stuff to get in. We needed space.

The Moore Building is famous because it housed the ENIAC all-electronic computer. The ENIAC hardware had occupied most of the first floor. After ENIAC was completed in 1945 and used successfully for two years, it was disassembled and moved to the U.S. Army sponsor's site at Aberdeen Proving Grounds in Maryland. Part of that now empty first floor of Moore was refurbished as a suite of TV teaching studios. The Moore School owned a radar research facility in nearby Valley Forge, Pennsylvania. At that time, the Valley Forge area was a local tech hub for a number of electronics and computing industries, including Univac, Burroughs, General Electric, and Lockheed-Martin. The Valley Forge site housed one of the remote teaching classrooms. Students could take Penn classes taught in the Moore Building with live outgoing video transmission and two-way audio. It was quite the technology after WWII.

A Moore School Electrical Engineering Professor, Octavio Salati, designed and operated this TV facility. Salati's problem was that he had no funding to upgrade his video systems after building it. My problem is that I had funds to buy new equipment, including a 3/4″ U-Matic video editing suite, but no place to keep it other than my office. By enabling portable equipment, the U-Matic tape cassettes freed video recording from expensive fixed studio requirements. I taught classes in one of Salati's TV classrooms, and so we struck up a friendship. I welcomed Salati's use of my video editor, as he had no 3/4″ tape recording equipment at all. It was a symbiotic relationship.

© The Author(s), under exclusive license to Springer Nature Switzerland AG 2025   59
N. Badler, *On Raising a Digital Human*, Synthesis Lectures on Computer Science,
https://doi.org/10.1007/978-3-031-63945-6_15

In the early 1980s, Salati retired from Penn. Anyone inside academia knows that space is *the* academic currency: more space, more prestige. Salati convinced Dean Joseph Bordogna, another electrical engineer, to cede the TV studio space and Salati's own large office on the first floor of Moore to CIS and place it all under my control. I had been recently tenured, which also helped. I moved into Salati's spacious office, and my cramped lab moved into a partially renovated, but still rather dingy, TV studio area. We called it the "Computer Graphics Research Lab." We became the new occupants of hallowed ground: the original ENIAC space was now ours.

Whenever I was in, I kept my office door open. It was just outside the lab entrance, so I could see who came in and out. A metal ramp led into our space from the hallway, and the cadence of footfalls announced the visitor. I usually succeeded in identifying which person came down the ramp leading to the lab and my office before they were visible. There were some downsides, though. One day, I learned that there was asbestos in the ceiling, and a remediation crew would be in to remove it. I never saw any asbestos up to that point but started to find it on bookshelves and counters after the remediation was supposedly complete. I won't discuss further the mouse problem. Nonetheless, overall, it was a great place to expand a research enterprise.

This Ex-ENIAC TV studio space included high ceilings, tall windows, and interior blank walls. We had room for numerous computer terminals and, eventually, Silicon Graphics workstations. We retired our Lexidata display. We took it apart and found that it contained numerous large, beautifully laid out printed circuit boards and integrated circuit chips. They were so attractive that we hung most of them on the lab walls as Tech Art.

Other activities in the CIS department also helped boost our research funding. In 1984, Joshi, Bajcsy, and I were the recipients of a major U.S. Army grant for a multi-year effort in "Artificial Intelligence Research." This gave us considerable freedom to support Ph.D. students, buy new equipment, and generally pursue our respective research programs under the AI umbrella. As a condition of the grant, some of us gave on-site tutorials on "AI" for the Army. I took the opportunity to proselytize our computer graphics, too. My association with the Army sponsors focused on their human factors interests in the Aberdeen Human Engineering Laboratory. Clearly, our TEMPUS work influenced the Army decision to fund me through the AI program, but I also received many subsequent individual grants through more specific human factors programs led by Ben Cummings, Brenda Thein, Bernie Corona, and Rick Kozycki at Aberdeen. Later, we added support from Steve Paquette at Army Natick Labs and from Art Iverson and Jack Jones of the Army TACOM Tank Command. I saw Army posters of TEMPUS figures inside a Bradley Fighting vehicle. We were accumulating real users.

In 1987, we added NASA Ames to our grant portfolio. The Ames group centered on aviation, especially helicopters. Beyond mere human fit in the cockpit, the Ames researchers were analyzing the pilot's mission-dependent physical, visual, auditory, and cognitive workloads. Our view of human factors broadened considerably with the addition of these active contextual dimensions.

There were some interesting twists to this. I had (naively) thought that because we were working with NASA Johnson Space Center, other NASA sites would be aware of our work. Not so. I was determined to bring the rest of the NASA human factors community together so we could work with all of them. I helped broker a face-to-face meeting across all the relevant NASA sites to discuss collaborative efforts in human factors stemming from our TEMPUS work. Representatives met and, to my great surprise, had to introduce themselves to their counterparts from other NASA locations. They had not met under a common "umbrella" before. I amused myself by thinking my epitaph would be: "He got all the NASA sites to talk to one another." I do believe this meeting had a positive collaborative outcome for everyone. NASA Ames soon became one of our strong supporters.

With ample research funds and generous space in the Computer Graphics Research Lab, we were able to purchase one of the new Silicon Graphics 1400 Workstations. I had already met Silicon Graphics founder Professor James Clark in 1981 when we shared a speaking venue at the Second European Computer Graphics Conference and Exhibition (Eurographics) in Darmstadt, Germany. Ginny, our almost two-year-old son David, and Ginny's sister, Adelaide, accompanied me on this visit. Clark brought along his eleven-year-old daughter, Kathy. Rather than letting Kathy wander around the city on her own, Ginny invited her to accompany them on a day trip to Frankfort. Oh, yes, and in 2000, Kathy Clark married Chad Hurley, co-founder of YouTube.

We looked forward to working with our new SGI workstation. Because SGI had a built-in depth-buffer rendering engine, we could concentrate on modeling and interaction in 3D. Decent funding also helped me attract more graduate students. One particular individual stood out, Ph.D. student Cary Phillips. I essentially dedicated this workstation for Phillips' use and let him explore it on his own. Soon I realized that my programming days were over. Phillips was just so efficient that there was no need for me to try programming anymore. Phillips also had very high code quality standards of his own. If I suggested that he incorporate some interesting piece of code from another student, as likely as not he would recode and improve it to his own satisfaction. He wasn't afraid to reject poor ideas or bad code, either.

Phillips was especially interested in how one might interact in 3D using only the keyboard and a 3-button mouse. The contemporary user interaction techniques espoused by Eric Bier of Xerox PARC in his paper "Skitters and Jacks" [1] influenced Philips' own approach. Bier's 3D cursor location was portrayed with a small "jack"-shaped icon of three short mutually perpendicular line segments that "stuck to" and "skittered along"

© The Author(s), under exclusive license to Springer Nature Switzerland AG 2025
N. Badler, *On Raising a Digital Human*, Synthesis Lectures on Computer Science,
https://doi.org/10.1007/978-3-031-63945-6_16

Fig. 16.1 Jill figure in a
NASA experiment glovebox

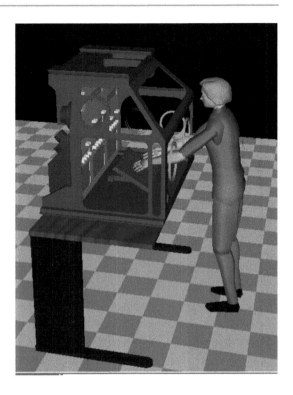

object surfaces and edges for selection and editing. The cursor looked like a piece from
the classic kids' game of jacks. Phillips adopted this cursor style and created his own
user interface control methodology. He named his program *Jack* [2]. The three mouse
buttons could be used to select X, Y, or Z dimensions, even simultaneously. For example, holding both X and Y buttons would move the cursor in the XY plane. Together
with keyboard keys, the mouse and buttons initiated global or local translations or rotations within a joint hierarchy. When applied to an articulated human model, interactive
posing was easy. Ph.D. student Jianmin Zhao's inverse kinematics code provided an interactive capability that allowed a hand or foot to reach any given input 3D cursor position.
The algorithm preserved segment integrity and respected joint limits [3]. Soon the name
Jack became synonymous with the human figure itself, and provided convenient and
memorable branding.

The early *Jack* human lacked a face and eyes, so Phillips modeled a visor cap for
the model's head to visually track its front-facing direction. The cap became a distinctive
feature for *Jack*. We did eventually add a face, since a cap was not always appropriate environmental attire. At some point, a female "Jill" model brought gender equality
(Fig. 16.1). Since not all programmers are artistically skilled, we welcomed art students
into the lab—an arrangement that would crystallize later as the Digital Media Design
undergraduate degree in Engineering at Penn.

I began to realize the research advantages of working with people who had to solve *real* problems. Our multi-year experience with TEMPUS gave us fantastic partners in the Crew Station Design Section at NASA JSC. Key staff included Barbara Woolford, Geri Brown, Jim Maida, Linda Orr, and Abhilash Pandya. My research team at Penn had to learn how to create an anthropometrically scalable human model [4, 5]. Our anthropometry modeling effort led to long associations with the human shape analysis community, including John Roebuck, Bruce Bradtmiller, and Kathleen Robinette. Collaborations with anthropometry experts incentivized us to develop accurate skeletal articulations for human models to predict postures able to navigate tight, obstructed, and awkward spaces.

Jack replaced TEMPUS as high-performance Silicon Graphics workstations became readily available, flexible, and affordable. Since our NASA connections had to evaluate complex International Space Station (ISS) configurations, *Jack's* human modeling had to improve. One default posture for the figure was in fact the zero-gravity pose, where all body muscles exert equal torques on a joint and result in specific joint bends, similar to a seated pose with out-turned thighs. With the deployment of the ISS, zero-gravity maneuvering and safe free-body locomotion would be critical design assessments. Later, with colleague Professor Dimitris Metaxas, postdoc Ambarish Goswami, and Ph.D. students Gang Huang and Suejung Huh, we would use physics-based dynamics simulations to reproduce motions in zero gravity [6].

Other modeling improvements in *Jack* included an accurate articulated spine model constructed by Master's student Gary Monheit [7] (Fig. 16.2). Rotational complexities in the clavicle and shoulder joints further motivated the development of accurate models for reach studies by Ph.D. student Deepak Tolani [8]. The NASA Ames MIDAS project—modeling helicopter pilot workloads—required foveal and peripheral visual field assessments. Based on input from Ares Arditi of The Lighthouse, a human vision clinic and research institute in New York City, we gave *Jack* an enhanced visual field model. Supported by Ames staff, including Kevin Corker, Barry Smith, and Brian Gore, we received especially good press in Gary Stix's 1991 column in Scientific American [9], which illustrated *Jack's* visual view capabilities.

John Deere human factors engineer Jerry Duncan, an early and enthusiastic *Jack* customer, provided heavy equipment use cases. For example, *Jack* could lean back and twist to look backwards over his shoulder to see what was happening behind the machine. Duncan also suggested a virtual mirror capability in the 3D models. Knowing *Jack's* eye position, we reflected it around the mirror plane, rendered an image from that virtual vantage point, and then mapped and clipped that image into the mirror polygon. The Silicon Graphics workstation had sufficient rendering capability to include at least a couple of mirrors in real time.

As early Silicon Graphics workstation adopters, we seemed to help market this hardware. *Jack* on an SGI showcased its best real-time graphics capabilities and had utility as well. I'm certain we motivated some of our potential *Jack* customers to buy SGI systems. An SGI was a worthwhile purchase for 3D designers, but *Jack* offered an additional

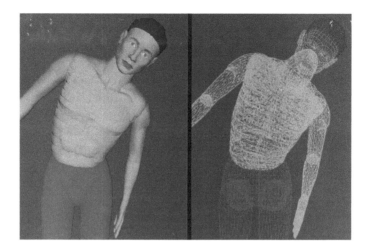

Fig. 16.2 Jack has a full anthropometric spine with vertebra and a segmented torso to show deformations

turnkey solution for a community of human factors engineers and analysts. The trick was linking a customer's 3D geometry CAD files to *Jack*'s internal mesh representation. I had met Bill Kovacs of Robert Abel and Associates through SIGGRAPH Conferences. Bill was a founder of Wavefront Technologies in Santa Barbara, California, one of the earliest computer graphics software companies. Through Kovacs, Wavefront donated its Advanced Visualizer software to my lab in 1989, and we installed it on our six SGI Iris 4D workstations. We were also designated a member of their "National Academic Advisory Committee." Most significantly, Wavefront developed and promoted the "OBJ" 3D mesh file format. OBJ files quickly became the de facto geometry transfer standard. We built an OBJ geometry converter to bring 3D CAD models into the *Jack* workspace. As long as a customer's CAD system could output geometry in the OBJ format, *Jack* could import it. Since all major CAD systems soon exported Wavefront OBJ files, *Jack* was an easy add-on. For practical reasons, users sometimes decimated unnecessary details from models for human factors analysis, and they had to position the human models interactively. On the whole, *Jack* was easily adopted into a company's workflow as long as they had an SGI workstation. Installation was straightforward, we had a well-written user manual, and students were available to assist and answer questions.

As a university lab, we could also offer a certain agility in providing new features. The research grant route was too awkward for most corporate customers, so we asked Penn's legal team to develop a license. Funds could flow into the lab through the license and provide additional support for student help. Our customers were loyal and expected active support, so they did not balk at yearly maintenance fees in return for reliable technical help and software upgrades. This regular revenue stream allowed us to purchase additional SGI workstations and, in the mid-1990s, an SGI "Reality Engine." This beast was the size of a refrigerator. The fans were so powerful that we built a "hat" to channel the airflow upward, away from the surrounding workstations. It was the lab "centerpiece" for many years until it was decommissioned in the early 2000s. We donated it to some students who carted it away and likely converted it into a storage cabinet. The only surviving evidence of our SGI era is a "Reality Tour" leather jacket that I won in a quiz game at a graphics conference.

Developing the *Jack* system meant on-the-job management training. Although I had led a small programming team at Kramer Research, I was not in charge. A few Ph.D. students capably organized our TEMPUS development. Since *Jack* was highly dependent on Phillips' programming skills, I found a way to get him more salary than the small, mandated Ph.D. student stipend. Phillips could then effectively function as my lab manager and separate me from lower-level software development tasks. Although faculty controlled Ph.D. appointments, subject to available research funds, the Engineering School controlled staff positions. I requested a "Systems Programmer" position for the Graphics lab. It was approved, Phillips applied, and I hired him. Somewhat surprisingly, this arrangement did not cost much more than having him as a Ph.D. student because research grants had to pay stipend, overhead, *and* university tuition. As staff, I only had to pay his much larger salary, employee benefits, and overhead. These effectively summed to similar totals. The bonus: as staff, Phillips had an employee benefit that gave him *free* tuition to complete his degree. This worked well all around. During the *Jack* development phases in the early 1990s, I added more Systems Programmer positions to the lab, employing Mike Hollick, Tripp Becket, John Granieri, and Pei-Hwa Ho.

Phillips finished his Ph.D. in 1991. During one of my more productive academic sabbaticals, I took his thesis as the backbone for my first book. Phillips and Professor Bonnie Webber were my co-authors [10]. This volume was the first computer graphics book to describe interactive human body models in detail. Other Ph.D. students contributed chapters on related graphics and natural language topics. Although I can't claim it was a best seller, the publisher (Oxford University Press) did sell out their print run. At my request, they relinquished the copyright, and the book now resides with free access on my website.

Phillips went on to an illustrious career in computer animation. After graduating, he joined Pacific Data Images. Soon after, he moved to Industrial Light and Magic (ILM) and became their lead creature animation engineer. At ILM, he received *three* Motion Picture Academy Technical Achievement Awards: the "Tech Oscars." Recently, I had an opportunity to ask Phillips why he chose Penn to do his Ph.D. He wasn't local to Philadelphia, and my reputation in the 1980s wasn't nearly as established as other computer graphics powerhouses such as Cornell, Brown, Cal Tech, and the University of North Carolina. Phillips had attended Johns Hopkins University as an undergraduate, and my previous Ph.D. student Professor Joseph O'Rourke was his computer graphics instructor. When Phillips asked for advice on Ph.D. programs, O'Rourke simply told him to "go to Penn and study with Badler." What a great and generous recommendation, and one that Phillips accepted to everyone's benefit.

Jack development continued in earnest between 1988 and 1996. One of my opening lines in any talk I gave was "At the Penn Computer Graphics Research Lab we divide the human body into graduate students." We had students working on faces, hands, spines, shoulders, legs, arms, and even innards. We had developed the premier *interactive* human factors modeling and analysis software system of its time. Its fame was spreading among the ergonomics community through talks, papers, publicity, and word-of-mouth. *Jack* opened doors I hadn't known existed. I was often the only computer scientist, let alone computer graphics person, in the room. I was invited to some meetings that were internal corporate retreats for the ergonomics and maintenance engineers and were normally closed to anyone outside the company. I often brought along a *Jack* programmer, usually Mike Hollick, to give a live demo. I rarely crossed paths with other academics I knew from the major SIGGRAPH conference. I did see some crossover the other way, though, where some practitioners in the human factors community we had helped expose to computer graphics were now curious enough to attend SIGGRAPH.

Air travel had its usual complications with delays and connections, but sometimes I got lucky. Mike Hollick and I flew to NASA JSC for a meeting. On the return, we departed through Houston Hobby airport. Besides being close to JSC, using Hobby avoided the downtown freeway traffic we would have to navigate to get to Houston International Airport. Because there were no nonstop flights to Philadelphia from Hobby, the main downside was having to make a connecting flight. On this particular trip, Hollick and I were bumped up to First class on a flight to Memphis, Tennessee. I sat next to a fellow with an unusually long beard even for the 1990s. He introduced himself: Billy Gibbons of the band ZZ Top. Being pretty oblivious to rock culture, I just introduced myself right back. I had no idea who he was or that he was the lead of a well-regarded rock trio. We talked during the flight. He was open about being a parent and the particular challenges that created. Hollick sat next to the group's drummer, Frank Beard, and was clearly more impressed with his good fortune.

I was promoted to Full Professor in 1987. Three years later, I became Penn CIS Department Chair. One of my first acts involved formalizing my lab as an Engineering School Research Center. The Computer Graphics Research Lab became the "Center for Human Modeling and Simulation" (HMS), which better reflected our mission, focus, and value [11]. Dean Joseph Bordogna approved the designation. I was now (besides department Chair) the Director of an official Penn "Research Center." The name transformation planted a proverbial "flag in the ground" to mark our territory. Emergent Internet searches would find us, given any appropriate search terms. No other entity ought to adopt the same name. We did what our name said we did. In 1992, during my tenure as department Chair, the Center also received a much-needed faculty boost. CIS hired University of Toronto Professor Demetri Terzopoulos' newly minted Ph.D., Dimitris Metaxas, as an Assistant Professor. Although Metaxas quickly and independently developed his own research enterprise, he and I would mutually benefit from our excellent collaborations over the next decade.

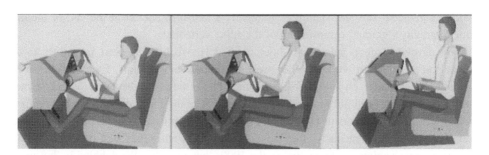

Fig. 16.3 Three different size *Jack* models in an automobile cockpit. The range of size variability from 5 to 95th percentile males is evident. (Note that the image scale is reduced in the right-hand figure, so that it will fit into the frame.)

With fame came jealousy. While some members of the human factors and ergonomics communities welcomed new interactive computer graphics tools such as *Jack*, not everyone was happy with us "upstarts." I remember meeting a senior human factors researcher in the stairwell at some conference, and he grumbled about me being a "dilettante." I think he missed the point of *Jack*: we didn't *do* human factors experiments and analyses; we just made tools to enable analysts to do their jobs. Different anthropometrically scaled bodies were essential to human fit and function analyses (Fig. 16.3). This point came home in another context when we were meeting with General Motors. They were interested in evaluating factory workcell and vehicle cockpit *comfort*. We added some features to provide suitable comfort measures [12]. They made it clear, however, that *Jack* was not to determine whether something *was* comfortable or not. We were to return numerical data to allow the analyst (or the management) to assess whether the action was indeed "comfortable" to their own corporate standard.

Obviously, we weren't trailblazers in human body mechanics, either. The field of biomechanics was robust and offered much-needed data to help create the *Jack* models. We relied on published studies, data, and body properties. We had neither the lab nor the expertise to undertake our own data collection. Segment mass distribution, centers of mass, and size came from anthropometry sources. Joint motions were trickier. There are several reasons why digital anthropometry is nontrivial. While external body measurements seem easy, they are based on external landmarks and mostly linear or circumferential measurements. On a digital model, what one really wants are internal joint centers of rotation. These have to be inferred from external points, as they cannot be measured readily without 3D medical imaging devices. The joints themselves are not always simple; shoulders, spines, and knees are particularly complex due to their multiple degrees of freedom. Body segments in between joints are not simple geometric shapes. Measuring the circumference of the chest, for example, does not capture nuances of its non-cylindrical, cross-sectional shape. Even a measurement as easy as full body height is not a guaranteed constant: stature is nominally a half-inch shorter at the end of the day due to gravity's gradual compression of the spine vertebrae.

We developed good spine and shoulder models. We had no overt incentives to do better knees, although we knew they were not simple hinge joints. I was invited to be an external evaluator on a U.K. biomechanics Ph.D. thesis on the knee joint. I learned much about what I did not know! I also paid a visit to the biomechanics lab at Stanford, where I met with Dr. Parvati Dev, a hand specialist. From her, I learned that finger joints don't all bend simultaneously; they have phased starts when flexed. I don't know how many computer animators knew that in the 1990s, but *Jack* did. We dabbled in, but made no contributions to, muscle-based skeletal movement. We were happy to leave that space to Stanford Professor Scott Delp and his OpenSim software [13]. I came to believe the major academic point of separation between biomechanics and animation is often the desire of biomechanics to characterize the nominal behavior of human motions, such as gait, averaged over numerous samples or individuals, while animation seeks the opposite by striving for and relishing individualization and uniqueness. This is not a criticism of biomechanics; we just have differing missions. We used motion capture systems for animation data collection, especially of gait, and the motion signature of the performer was often visibly detectable in the results.

Unlike some other proprietary human models, *Jack* was not bound to any particular manufacturer's CAD system. Geometry converters let *Jack* and *Jill* inhabit any 3D imported mesh model (Fig. 16.3). Being agnostic to the computer environment (as long as one had a Silicon Graphics workstation) was a huge advantage because we could license *Jack* to nearly anyone. As *Jack*'s capabilities grew, we also obtained funding from industry partners, including FMC, Siemens, Kimberly Clark, General Motors, Lockheed-Martin, Martin-Marietta Denver Aerospace, Litton Data Systems, and Deere. An unexpected obstacle appeared as we sought funding from numerous U.S. government agencies: we were asked why government branch X should also support work that government branch Y was funding. This even had a name: the silo problem. Our response was simply that they get what they fund *plus* all the features that everyone else funds. That nailed it. We were able to receive support for *Jack* from the U.S. Army, U.S. Navy, U.S. Air Force, NASA JSC, NASA Ames, NASA Goddard, NASA Kennedy, Battelle Pacific Northwest Labs, Defense Modeling and Simulation Office, ARPA, NIST, and the National Library of Medicine. Everyone received the same unified version.

With so many disparate customers and applications, one of my Ph.D. students, Welton "Tripp" Becket, had a brilliant idea. He wrote an application programming interface (API) in the Lisp language for *Jack*, so that all its functionality could be accessed and even extended through external software [14]. *Jack* was now not only the premier interactive digital human system, but it was also *embeddable* into other applications. We called this extension the "*Jack* Toolkit." Among the most significant uses of the *Jack* Toolkit was our animated conversation work (which I will describe later), where two *Jack* models were simultaneously controlled by AI planning, gesture, and speech generation systems.

Having the API in Lisp was not a real restriction, and eventually it was rewritten in C++ by another Ph.D. student of mine, Dr. Paul Diefenbach. The C++ API was the backbone of a later development called DEPTH, which we undertook for the U.S. Air Force. But that's a story in itself.

Between *Jack's* usable software, our recognition of anthropometric variability, and the publication of our *Simulating Humans* book, we were recognized as a potential *International Standard* for digital humans. In the early 1990s, the *Jack* body became the foundational skeletal structure model for the MPEG-4 standard *H-Anim* [15]:

> This International Standard specifies H-Anim, an abstract representation for modeling three-dimensional human figures. This International Standard describes a standard way of representing humanoids that, when followed, will allow human figures created with modeling tools from one vendor to be animated using motion capture data and animation tools from another vendor.

The Bibliography for this Standard includes our *Simulating Humans* book as its only academic citation. The present ISO standard is ISO/IEC 19,774–1:2019. This effort is still actively supported by a dedicated community at h-anim@web3d.org/.

What ultimately surprised me the most with the *Jack* software was the impact it had on the engineers and analysts who used it. We usually had no access to the projects they were doing. However, because we were always close to the actual users and not insulated through impersonal procurement or management decisions, we could gauge their reactions for ourselves. NASA's Jim Maida, who began working with us in the TEMPUS days, had switched to *Jack* and used it regularly for Shuttle and ISS applications. I saved a personal letter from Ed Bellandi of FMC who told us how *Jack* had given his human factors career meaning and satisfaction. I experienced these emotional responses firsthand on a "marketing" visit to the aircraft manufacturer Embraer in Sao Paulo, Brazil. I sat with the human factors engineer who had been using our usual 30-day free *Jack* trial license. The license had just expired, and Embraer was considering a purchase. This poor fellow, who had been the *Jack* user for a month, was tearfully lamenting that he was now obligated to return to using AutoCAD to perform his analyses. Embraer soon adopted *Jack*.

Over the latter part of my career, I sometimes consulted as an expert witness on patent cases. In 2014, I was hired (almost last minute) as the animation expert on a case being tried in federal court in Austin, Texas. I had never been to Austin, though I had traveled elsewhere in Texas many times, especially Houston, due to my long association with NASA JSC. During the trial, I sat with the other experts and observers on the bench seats at the rear of the courtroom. The day it began, a woman came in and sat in the front row of benches. She spoke with one of the attorneys whom I was working with, and I learned that she was his wife. "That's nice," I thought, but paid little attention to the situation otherwise. She attended the trial again the next day. At the end of that day, another expert offered to drive people back to the hotel: it was June in Austin and hot. He also asked the woman if she wanted a ride as well. She agreed. We ended up sitting together in the back seat of his car. I asked her if she always watched the cases her husband presented in court. "Oh no," she said, "The last time I went to observe was nine years ago." So this trial was special and her attendance was most unusual. She went on to remark how much she liked Austin over Houston. Apparently, she used to live in Houston. In fact, her husband used to work for NASA in Houston. And, yes, she had worked at NASA, too. I mentioned my long-standing research association with the Crew Station Design Section there and asked if she knew any of these folks. Well, not only did she know them all, she had shared her office with my primary research connection there, Barbara Woolford!

References

1. E. Bier. "Skitters and jacks: Interactive 3D positioning tools." Proc. of the 1986 workshop on Interactive 3D graphics (I3D), Jan. 1987, pp. 183–196.
2. C. Phillips and N. Badler. "*Jack*: A toolkit for manipulating articulated figures." ACM/SIGGRAPH Symposium on User Interface Software, Banff, Canada, October 1988, pp. 221–229.
3. J. Zhao and N. Badler. "Inverse kinematics positioning using nonlinear programming for highly articulated figures." ACM Transactions on Graphics 13(4), Oct. 1994, pp. 313–336.
4. M. Grosso, R. Quach, E. Otani, J. Zhao, S. Wei. P.-H. Ho, J. Lu and N. I. Badler. "Anthropometry for computer graphics human figures." Technical Report MS-CIS-89–71, Department of Computer and Information Science, University of Pennsylvania, 1989.
5. M. Grosso, R. Quach and N. Badler. "Anthropometry for computer animated human figures." In *State-of-the Art in Computer Animation*, N. Magnenat-Thalmann and D. Thalmann (Eds.), Springer-Verlag, 1989, pp. 83–96.
6. G. Huang, S. Huh, A. Goswami, D. Metaxas and N. Badler. "Dynamic simulation for zero-gravity activities." International Space Human Factors Workshop, Tokyo, Japan, June 1999.
7. G. Monheit and N. Badler. "A kinematic model of the human spine and torso." IEEE Computer Graphics and Applications 11(2), March 1991, pp. 29–38.
8. D. Tolani and N. Badler. "Real time human arm inverse kinematics." Presence 5(4), 1996, pp. 393–401.
9. G. Stix. "Human Spec Sheet." Scientific American, November 1991, p. 132–133.
10. N. Badler, C. Phillips and B. Webber. *Simulating Humans: Computer Graphics, Animation, and Control*. Oxford Univ. Press, 1993. https://www.cis.upenn.edu/~badler/book/book.html
11. N. Badler, D. Metaxas, B. Webber and M. Steedman. "The Center for Human Modeling and Simulation." Presence 4(1), 1995, pp. 81–96.
12. P. Lee, N. Badler, S. Wei and J. Zhao. "Strength guided motion." ACM SIGGRAPH Computer Graphics 24(4), 1990, pp. 253–262.
13. OpenSim. https://simtk.org/projects/opensim/ . Accessed April 26, 2024.
14. W. Becket. "The Lisp API Version 1.1." University of Pennsylvania Computer and Information Science Technical Report, 1994.
15. https://www.iso.org/standard/64788.html . Accessed August 26, 2023.

Jack Grows Up

17

By 1994, I felt as if I was running a small business within the university. At the peak of activity, I had 24 Ph.D. students, four full-time staff programmers, international licensing agents in the U.K. and Israel, multiple licensed sites, ongoing software maintenance fee income, and a mailing list in the hundreds. Karen Carter, my full-time administrative assistant, kept me and the *Jack* community organized. I juggled so many sponsored projects that I briefly adopted Microsoft Project® to keep track of personnel tasking, proposal due dates, reporting deadlines, and how funds were distributed. I abandoned that approach when I realized it would be almost another full-time job just to keep the Project file current.

During this period, I discovered my personal "gift": I was "infinitely interruptible." Students and staff had ready access to me, and I would stop writing or whatever and handle the issue. Thank goodness for capable staff and self-motivated students. I learned that I saved time by managing issues on the spot rather than taking up time to schedule a future meeting. We joked about getting me a "Take-a-number" system to maintain order. We bought benches for the foyer outside my door to provide a more comfortable wait.

I became "salesman *Jack*" and raked up considerable travel to show and license the system to new users. One year I even handmade a cardboard *Jack* costume for the lab Halloween party (Fig. 17.1). *Jack* became a generic brand (just like Kleenex was a generic brand for any tissue): I had various people tell me they saw *Jack* being used when in fact I knew that they were talking about someone else's human model. This success was both great and terrible. The time was ripe to move *Jack* out of the university and into its own company.

The spin-off process began with Penn's Center for Technology Transfer. They had been involved since the early 1990s and had offered to license all the *Jack* intellectual property

© The Author(s), under exclusive license to Springer Nature Switzerland AG 2025
N. Badler, *On Raising a Digital Human*, Synthesis Lectures on Computer Science,
https://doi.org/10.1007/978-3-031-63945-6_17

Fig. 17.1 The author in his homemade *Jack* Halloween costume

to a robotics company for about $60K. I was relieved when I found out that the company rejected the offer. That deal had been floated without my knowledge. I felt that *Jack* was worth far more. That robotics company went on to build its own human modeling system, but it was locked into its existing software environment. Upon my urging, the Center for Technology Transfer did help trademark the name *Jack*, to at least minimally protect Penn's intellectual property, but we never initiated any patent claims. I also give Penn full credit for generating the appropriate legal documents for our *Jack* licensing activities.

A change of staff in the Technology Transfer office brought in Lou Berneman. He proposed creating a spin-off company rather than selling the *Jack* intellectual property outright. Working through a local venture capital firm, Safeguard Scientifics, we realized that the window to make a deal was starting to close. We were not secret about our algorithms, and we gave out free trial licenses, so it would not take long for a powerful corporate interest to reproduce a *Jack*-like system on their own. We already knew of at least one company that was actively engaged in making a *Jack* "clone" based on PCs rather than SGIs, because we kept running into them and their demos at the same trade shows and meetings.

The process of generating a spin-off company in the *software* space was something never before attempted by Penn's Center for Technology Transfer. Every step of the process was novel; every aspect of the eventual licensing agreement had to be generated from scratch. The Safeguard folks and Penn staff knew that the spin-off could only happen if they could find an able CEO to run it. I quickly realized that being a CEO would be different from being a university faculty member, so I did not want to be a candidate. In fact, I really wanted to disengage completely. I was longing to get back to being a researcher and teacher, and retire from the salesman role.

Penn wisely retained a professional search firm to identify potential CEO candidates. Within a few months, they zeroed in on an MBA from the University of Michigan, Jim Price. We arranged a *Jack* user meeting at Penn and invited all our sponsors and licensees. It was not the first time we had done such a user meeting. This one's real purpose, however, was to give Price a first-hand opportunity to converse with our customers, see their applications, and receive unvarnished feedback. By the end of the day, Price was convinced that *Jack* had "legs." Price agreed to become the CEO of a new company he would form called "Transom Technologies."

Starting a company requires capital. The Safeguard folks set a goal to raise $3M to launch Transom. To avoid unequal ownership, they required that any investor had to put up exactly $100K. We set up pitch meetings with likely investors. The usual format included short talks by Price and me, and a live demo if possible. Then, the Safeguard representative would give his pitch, and investors were invited to line up and write their checks. At the third pitch session we reached the $3M mark. The line still had eager investors, but they were turned away when we met our target.

As CEO, Price wanted to maintain the Transom operation in his hometown of Ann Arbor, Michigan, but all the knowledgeable *Jack* programmers were at Penn. He opened a satellite office in Philadelphia and hired several of my ex-student staff members to maintain continuity. For a couple of years Transom kept offices in both Ann Arbor and Philadelphia. The choice to keep the main office in Ann Arbor was ultimately a smart one because there was a strong local ergonomics program founded by Professor Don Chaffin at the University of Michigan. Price hired Michigan Ph.D. Dr. Ulrich Raschke to join Transom. *Jack* finally had an "authentic" human factors researcher on staff.

The 1996 *Jack* move to Transom left me in a psychological limbo. I was happy that *Jack* marketing and support was no longer my problem. But, as it was the HMS Center's tangible product for eight years, it was like seeing a child go off on their own. Price now took responsibility for *Jack*'s continued success. As CEO, he was entitled to make decisions that appeared averse to the structure I had built, but better aligned to his corporate vision. I think he made the right decision by not listening to me advocate for four divisions within Transom: ergonomics, games, medical, and AI agents. He chose wisely to focus on just the first of these. Some of his other choices were emotionally difficult for me. For example, he withdrew Transom from our overseas reseller agreements because he did not like the terms. Our U.K. agent GMS had been especially effective in promoting *Jack* in Europe. They printed beautiful full-color brochures and even built a training center to support customers.

Early on, Price hired a non-technical sales manager but soon realized that the most effective salespeople were those who could explain and use the technology themselves. In my opinion, the overall best decision he made (other than hiring my former students!) was to quickly develop, program, and distribute a PC-compatible *Jack* version. We did not have the resources for such a non-research software conversion at Penn. Moreover, the mid-1990s were transitional for computer graphics hardware. Silicon Graphics was waning while GPU boards slotted into PCs were rapidly taking over the computer graphics market. Overall, Price was the force I needed to reset me into my role as a researcher and educator. I will be eternally grateful to him for enabling that transformation.

As part of the *Jack* spin-off license agreement, the core developers received shares in Transom. The license (deliberately) did not include any royalty provisions, so our shares would only acquire value if Transom was sold to a publicly traded company or had its own IPO. In 1998, Engineering Animation (EAI) bought Transom. EAI had a popular following as a primary source of pre-built 3D polygon mesh models of many common objects—such as vehicles, furniture, and buildings—that would be more costly for individuals to create on their own (at that time!). The Transom shares were vested and now worth real money. Unfortunately, EAI had a major internal issue within a year that caused its share price to drop precipitously, to about a sixth of its former valuation. I could honestly say, "I used to be a millionaire."

With its reduced valuation, EAI was now itself an acquisition target. A company called Unigraphics purchased EAI, automatically cashing out all EAI shares. We did put our windfall to good use, though, as a substantial downpayment on a second home in upstate New York that had been built by Ginny's aunt and uncle after WWII. We were able to keep that house in the family when it was in danger of being sold to strangers.

Penn also received its share of the stock funds as part of the licensing agreement. The Center for Technology Transfer knew that the amount of money going to the university was rather insignificant with respect to its cash flow, but it was still substantial to us. The Technology Transfer folks agreed to return the university share to the developers, that is, to the *students* who contributed to *Jack* over the years. Phillips and I created a list of every student who contributed to the *Jack* software and binned them into high, medium, and low levels of involvement. The university proceeds were divided proportionally among these sets, and every student received a decent cash payment from Penn. I'm sure most of them never expected such a return. The goodwill was the best part of the deal. I could also imagine my former mentor, Henry Kramer, smiling with approval.

Soon UGS, a General Motors software subsidiary, bought Unigraphics. *Jack* began to appear in the software portfolio that GM licensed to Engineering schools. Not long after, Siemens bought UGS and integrated *Jack* with its Product Lifecycle Management tools. *Jack* is currently used extensively for large-scale factory and workplace design and evaluation. Another company that Siemens purchased, Technomatix, contributed additional components to *Jack*. Perhaps not so coincidentally, Technomatix had itself tried to acquire *Jack* from Penn prior to 1996, but the Center for Technology Transfer turned down the deal. *Technomatix Jack* is now its official name at Siemens. Over 25 years after leaving Penn, *Technomatix Jack* is available as a free download for educational use [1]. There is one last twist in this progression. In 1991, I went to Munich, Germany, to give a talk on *Jack* at the Siemens Corporate Research Center. They had kindly given me a year-long corporate grant for some new *Jack* components. Although they expressed modest interest then, I had no idea that a decade later they would own it.

One of my original student staff programmers, Mike Hollick, stayed on through all these corporate hops, working remotely from Philadelphia even after the Transom Philadelphia office closed. He is still with Siemens as the lead *Jack* programmer. Looking even further back in time, I realized that the name *Jack*, which Phillips adopted entirely on his own, now paid accidental homage to the person who gave me a start in computers: "Jack" Jackson.

Jack was certainly not the first virtual human model. Fetter's line models date from 1964. Other CAD-proprietary models were in use at aerospace and vehicle manufacturers. Ed Catmull and Frederic Parke at the University of Utah built solid polygon hands and faces in the 1970s [2]. Don Herbison-Evans constructed a "sausage woman" out of ellipsoids [3]. Jim Blinn built a "blobby man" to illustrate his method of smooth shape blends with potential functions [4]. Nadia Magnenat-Thalmann and Daniel Thalmann and colleagues were creating animated smooth mesh characters, including a virtual Marilyn Monroe [5]. Animation studios explored 3D characters, including Jeff Kleiser and Diana Walczak's 1988 "Nestor Sextone for President" [6]. What made *Jack* unique in the late 1980s was its grounding in the scientific communities of anthropometry and human factors, its on-screen interactivity, color graphics, its extensibility through an API, and its growing user community.

References

1. Siemens. https://resources.sw.siemens.com/en-US/download-tecnomatix-jack-student
2. https://en.wikipedia.org/wiki/A_Computer_Animated_Hand
3. D. Herbison-Evans. "Real-time animation of human figure drawings with hidden lines omitted." IEEE Computer Graphics and Applications, 2(9) September 1982.
4. J. Blinn. "A generalization of algebraic surface drawing." ACM Trans. Graphics, 1(3), Jul. 1982, pp. 135–256.
5. N. Badler. "Animation 2000++." *IEEE Computer Graphics and Applications* 20(1), Jan.-Feb. 2000, pp. 28–29.
6. https://www.youtube.com/watch?v=nzRluq7_45c

I haven't been able to recall definitively how I became interested in animating faces. By a process of eliminating positive incentives, it could simply have been the lack of facial features on Bubbleman. Or it could have been an expressed interest by student Steve Platt, who produced his Master's thesis on the topic in 1980. Platt constructed a linear segment model of a face "mask" (not a full head) and animated it by moving the vertices of the mesh structure. Platt soon became my Ph.D. student. We wondered whether there was any psychological basis for controlling face vertices, say, to animate emotions or speech. Being still somewhat new to Penn (and lacking Internet search!) I asked my department Chair, Aravind Joshi, if he had any suggestions about relevant faculty I could talk to in Psychology. He pointed me to Professor Paul Rozin. I walked over to Rozin's office, and we had a nice chat about what Platt and I were trying to do. He admitted he wasn't quite the right contact. He told me to check into new work on representing facial expressions by a University of California at San Francisco researcher, Paul Ekman. I wrote to Ekman, requesting a meeting during my next trip to California. He agreed, and I made my second pilgrimage to talk to the best person possible.

During my visit with Ekman, he described his Facial Action Coding System (FACS). I was already quite familiar with several human *skeletal* motion representations (such as Labanotation). I also knew about an animated face mesh with independently controlled mesh vertices developed by Fred Parke at the University of Utah. Ekman's FACS approach was the first time I saw an imminently digitizable notational system for general facial movements. FACS simply consisted of temporally-ordered muscle activations, called Action Units (AU), displacing key face skin points. This seemed ideal for Platt's face model (Fig. 18.1). Platt turned the face mesh into a spring-mass system, locating face nodes at key muscle insertion points. He then implemented a physics simulator for

© The Author(s), under exclusive license to Springer Nature Switzerland AG 2025   79
N. Badler, *On Raising a Digital Human*, Synthesis Lectures on Computer Science,
https://doi.org/10.1007/978-3-031-63945-6_18

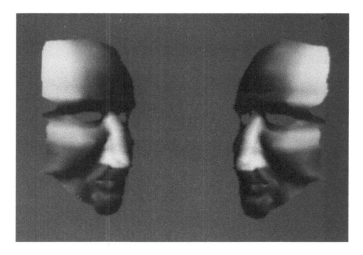

Fig. 18.1 Platt's face model rendered as shaded polygons

the spring motions. By contracting facial "muscles," the spring system deforms. Because the muscles are aligned with FACS AUs, Platt had realized the first computer graphics physics-driven face controlled by Ekman's FACS representation [1]. To avoid extreme deformations, such as springs moving too distant vertices, Platt's completed Ph.D. work explicitly bounded the spring influence areas. We sent our results to Ekman, who was pleased to hear of our success.

Platt's face work attracted the attention of people defining the emergent MPEG-4 codec standards. MPEG-4 was designed to aid in low-bandwidth multimodal data transmission, such as speech, object animation, and faces. Platt's use of Ekman's FACS AUs as discrete parameters to drive an animated 3D facial model fit perfectly into the MPEG-4 framework.

For a few years after Platt's 1985 Ph.D., I set face animation interests aside. Then, a remarkable Ph.D. student joined my lab at Penn, Catherine Pelachaud. Pelachaud picked up the face animation idea and began constructing a full face and head model with a smoothed polygon mesh. She also modeled eyes and lips, so the model could portray a wider variety of FACS AU movements as well as speech. She adopted the FACS-motivated MPEG-4 face animation representation as the animation controller. Pelachaud was the consummate Ph.D. student. She kept extensive notes in the same manner as I had while evolving my own Ph.D. Her notes were a delightful stream of consciousness mélange of French and English.

Pelachaud began to collaborate with Penn CIS Professor Mark Steedman on speech animation. Animation parameters could move lips to standard positions necessary to articulate any possible phoneme. Given a speech signal discretized into a temporal sequence of phonemes, each phoneme activated local mesh deformers that moved lip model vertices. This was a nice idea, and one can still see a similar process driving real-time avatars

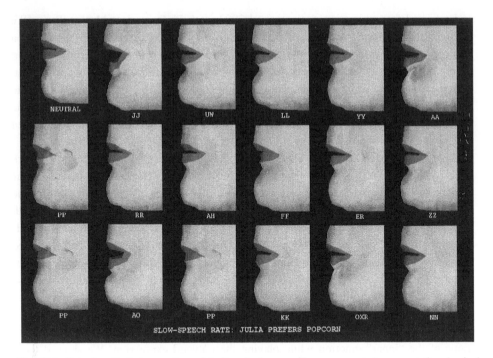

Fig. 18.2 Pelachaud's fast speech model altered phoneme lip movements to respect the physical abilities of the lip muscles. This process is called coarticulation

today. The problem, as Pelachaud discovered, is that's not quite what actually happens with lips. In reality, lip movements are constrained by physics: they can only move so fast. When phonemes occur in rapid succession, the lips do not have time to return to a neutral rest position. In fact, they may have to prepare to articulate the next phoneme before the current one is even finished. This effect is called *coarticulation*. In English, up to five successive phonemes may influence each other's lip performance. If an animation approach fails to handle coarticulation properly, the lips appear to flap or move in a vertical direction much too rapidly on fast speech (Fig. 18.2).

After receiving her Ph.D. in 1991, Pelachaud remained at Penn as a postdoc, mostly working with Steedman on the animation of speech and the synchronization of facial expressions and speech intonation [2]. Soon, another postdoc from France joined the group, Marie-Luce Viaud. We proposed to NSF and received funding for a seminal face workshop in 1994. The workshop invited a broad spectrum of face researchers from computer graphics, psychology, medicine, communications, and art. Penn's Institute for Cognitive Science hosted the event and published our Final Report [3]. Pelachaud returned to Europe to take a faculty position. She continued outstanding work in face animation, as well as providing open source software for her "Greta" face model. In 2015, Pelachaud was awarded the ACM/SIGAI Autonomous Agents Research Award.

I revisited face animation again around 2003, with Ph.D. student Meeran Byun. We observed that all face animation models at that time only concerned themselves with a more-or-less oval frontal facial mask. When we looked at professional actors, however, we noticed that many facial expressions actually engaged the mostly linear, but prominent, muscles in the neck. The stronger the emotional expression, the more obvious were the neck muscle bulges on the skin surface. Byun began to build this musculature into the Greta face model. Unfortunately, we could not complete this work and were discouraged to continue it with the publication of a similar approach by Demetri Terzopoulos [4].

After I thought we were out of the face animation business, I had a chance to work with a year-long Ph.D. student visitor from Brazil, Rossana Baptista Queiroz. She was obligated to spend a so-called "sandwich" year abroad as part of her requirement for the Ph.D. degree from the Pontifícia Universidade Católica do Rio Grande do Sul in Brazil. I already knew her supervisor, Professor Soraia Musse, from Daniel Thalmann's EPFL research group in Switzerland. We welcomed Queiroz to our lab. Queiroz and I decided to examine another aspect of face animation that had been missing in the computer graphics simulation literature: microexpressions. Paul Ekman introduced psychologists to these fleeting facial emotion displays as potential markers for lying. In general, microexpressions seemed to "leak" a person's true emotion while they were overtly expressing a different "macro" emotion. We arranged the Greta face simulation to swap in a single frame of the "inner" emotion while playing the outward expression. We showed these hybrid animations to human subjects. Generally, most of the participants recognized the foreground macro emotion, and most of the time, they perceived the subtle presence of the second, background, micro emotion [5]. Queiroz returned to Brazil with a nice publication from her visit and earned her Ph.D. soon thereafter.

References

1. S. Platt and N. Badler. "Animating facial expressions." ACM SIGGRAPH Computer Graphics 15(3), Aug. 1981, pp. 245–252.
2. C. Pelachaud, N. Badler and M. Steedman. "Generating facial expressions for speech." Cognitive Science 20(1), pp. 1–46, 1996.
3. C. Pelachaud, N. Badler and M.-L. Viaud. "Final Report to NSF of the Standards for Facial Animation Workshop." University of Pennsylvania Department of Computer and Information Science Technical Report No. MS-CIS-94-01, January 1994.
4. S. Lee and D. Terzopoulos. "Heads up! Biomechanical modeling and neuromuscular control of the neck." ACM Transactions on Graphics 25(3), August 2006, pp. 1188–1198.
5. R. Queiroz, S. Musse and N. Badler. "Investigating macro- and microexpressions in computer graphics animated faces." Presence 23(2), Spring 2014, pp. 191–208.

In the early 1980s, even as we were developing the TEMPUS software for NASA, the connections between graphics, animation, and natural language were still my fundamental motivating research thread. I imagined returning to the computer vision problems that dominated my Ph.D. thesis work. I even collaborated with a former Toronto Ph.D. student, Professor John Tsotsos of York University, to co-chair a workshop that produced an anthology spanning computer graphics and computer vision approaches to motion [1]. However, that was not to be my future.

Not long after I arrived at Penn, the CIS department hired a new Professor, Bonnie Webber, who had been working at Bolt Beranek and Newman (BBN). Webber had a strong reputation in natural language processing, particularly in the area of pronoun reference: "to what noun in an utterance does a pronoun refer?" Webber and Joshi were a good collaborative match in the natural language space.

Soon after Webber joined Penn, she (re-)introduced me to Professor Gabor Herman. Herman previously revolutionized medical imaging by creating an algorithm for turning the set of spiral-captured X-ray images from a computerized tomography (CT) scanner into a single coherent 3D voxel (volumetric) model. I knew about Herman's work while he was at the University of Buffalo, even before he came to Penn. I likely met him in person through his paper presented at the 1976 SIGGRAPH Conference in Philadelphia [2]. Herman's work had an enormous impact on the medical community. When I gave a talk on my research at the "International Forum of New Images Conference" in Monte Carlo in 1984, the lone medical doctor in the audience only wanted to talk to me afterward about Herman's work.

It is possible that my graphics connection with Herman drew me closer to a collaboration with Webber. Webber and I were both interested in interactive systems, but from contrasting perspectives: Webber in language and I in graphics. We created ab initio and co-taught undergraduate courses on interactive systems in 1980 and 1982. I welcomed Webber as a partner in our TEMPUS work, as my long-range vision was to have natural language input control the virtual human [3].

NASA's support for TEMPUS did not have an overt language component, although we explored the idea of animating operational instructions anyway [4]. Nonetheless, the notion that written instructions could drive digital humans intrigued me. Webber and I shared an existing research role in a broad interdisciplinary grant first sponsored by the Sloan Foundation and then formalized through the NSF as a Research Center in Cognitive Science. Computer graphics could play a funded role in the NSF Center, and interdisciplinary studies were encouraged. Webber and I agreed that an informal weekly seminar would be the best forum to engage Ph.D. students in a deep exploration of language and animation issues. We spent most of the first year trying to understand each other's perspectives and problems and identify a common, interesting, viable, and meaningful direction. Problems that were exciting in the language space were often too hard or less relevant to graphics, and vice versa. Late in the first year, we finally found our way: we named our seminar ANIML (Animation from Natural Language) and centered on instructions [5]. We would continue to collaborate on this topic until Webber left Penn in 1998, a run of almost twenty years.

By the late 1980s, I was firmly embedded in the computer graphics community. Both graphics and the ANIML seminar kept me aware of other approaches to human motion simulation. Professor Brian Barsky's UC Berkeley Ph.D. student, the late Jane Wilhelms, studied physics-based dynamics simulation of human figures. I also knew of Ph.D. work on skeletal control systems by David Zeltzer at Ohio State University. Barsky, Zeltzer, and I hatched a plan to hold an invitation-only workshop on human motion. In 1989, Zeltzer was on the MIT Media Lab faculty, so they hosted the workshop. The speakers covered the principal communities: computer graphics, robotics, mechanical engineering, psychology, and physiology. Each approached human animation from their own disciplinary perspective, so our intention was to help break down those barriers. The Keynote Speaker, Frank Thomas, presented classical Disney animation techniques and distributed autographed copies of his book to the attendees [6]. We had exceptional talent assembled for the Workshop, and the event would have disappeared into obscurity had the three organizers simply let it end. Barsky and Zeltzer advocated for a published volume of selected contributions from the invitees. Barsky, as Senior Editor of Morgan-Kaufmann's *Computer Graphics and Geometric Modeling* series, convinced publisher Mike Morgan to produce the volume. This seminal cross-disciplinary work debuted in 1990 [7].

References

1. N. Badler and J. Tsotsos (Eds.). *Motion: Representation and Perception*. Elsevier Science Publishers, North Holland, 1986.
2. R. Robb, E. Ritman, J. Greenleaf, R. Sturm, G. Herman, P. Chevalier, H. Liu and E. Wood. "Quantitative imaging of dynamic structure and function of the heart, lungs and circulation by computerized reconstruction and subtraction techniques." SIGGRAPH Conference Proceedings, 1976, pp. 246–256.
3. N. Badler, B. Webber, J. Korein and J. Korein. "TEMPUS: A system for the design and simulation of mobile agents in a workstation and task environment." Proc. Trends and Applications Conference, Washington, DC, 1983, pp. 263–269.
4. N. Badler and J. Gangel. "Natural Language Input for Human Task Animation." Second Annual Workshop on Robotics and Expert Systems, Instrument Society of America, Houston, TX, June 1986, pp. 137–148.
5. N. Badler, B. Webber, J. Kalita and J. Esakov. "Animation from Instructions." In *Making Them Move: Mechanics, Control, and Animation of Articulated Figures*. N. Badler, B. Barsky and D. Zeltzer (Eds.), Morgan-Kaufmann, 1990, pp. 51–93.
6. F. Thomas and O. Johnson. *The Illusion of Life: Disney Animation*. Abbeville, 1981.
7. N. Badler, B. Barsky and D. Zeltzer (Eds.). *Making Them Move: Mechanics, Control, and Animation of Articulated Figures*. Morgan-Kaufmann, 1990.

The ANIML seminar became a rich source of interesting problems and fruitful collaborations between natural language and graphics Ph.D. students. Numerous theses resulted, and Ph.D. student careers were launched across the ANIML content spectrum. Instructions were a tangible link between telling a virtual person what to do and having them actually perform the task on simulated geometry. Early on, we realized that language and graphics occupied rather distinct sectors of descriptive space:

	Natural language	Graphics
Geometry and shape	Awkward	Excellent
Action and behavior	Excellent	Awkward

For example, describing the furniture and layout of a room in natural language requires many words and still will not properly convey essential nuances (let alone numerical quantities) of shape, location, and size. On the other hand, computer graphics excels at allowing the creation of exact 3D models with mathematically precise size, shape, material, and position. Conversely, natural language is exquisitely poised to describe human actions and behaviors through verbs and adverbs and have them interpreted through personal experiences and performances of such actions. Computer animation can produce motions, but most of the classic techniques rely on the creation of mathematical functions to describe trajectories, joint angles, and explicit timings. Often, the methods required to form an animation are interactive, so that the process may be iterated until satisfactory. The ANIML conclusion is straightforward: exploit the best of both modalities by describing actions in natural language and interpreting the desired behaviors in the explicit 3D context of a graphically modeled space. Maintenance instructions were a perfect fit to this perspective.

© The Author(s), under exclusive license to Springer Nature Switzerland AG 2025   87
N. Badler, *On Raising a Digital Human*, Synthesis Lectures on Computer Science,
https://doi.org/10.1007/978-3-031-63945-6_20

Webber and I began collecting instruction examples from "how to" books, printed directives on part boxes, and instruction manuals from our sponsors, notably NASA and the U.S. Air Force. The examples ranged from the humorous and almost useless ("Replace windshield wiper with contents of this box") to the Air Force Technical Orders that had to be followed exactly or else the maintainer was subject to Court Martial! I collected these for a while since instructions were so easy to come by. There is an art to writing clear and correct instructions. We were amused with buggy ones, too. My favorite was in printed instructions for replacing a simple locking doorknob: after running through two pages of diagrams and text, the final instruction was "If your door is hinged on the other side, reverse Step 1."

One of our earliest attempts to drive a human model from instructions arose with TEMPUS. NASA produced extensive instructional and operational materials for astronauts. Space flight is complex, and manual equipment controls are essential for astronaut situational awareness and survival. Master's student Jeff Gangel connected NASA checklist procedures in simplified language syntax to panel control locations. TEMPUS could then animate simulated eye gaze directions and hand reaches [1]. For example, Ph.D. students Jeff Esakov and Moon Jung had two TEMPUS human models appearing to activate Shuttle Remote Manipulator System controls directly from a checklist [2]. We learned an important lesson from this exploration, namely, that instructions usually don't indicate to the animation how much *time* a task will take. Instructions just expect that the human will do the best and most accurate job they can. Our animation system needed to know the task *duration* in advance. We found that there was a psychophysical basis for establishing task timing: Fitts' Law [3]. In essence, Fitts' Law produces a decent estimate of task time given the size of the target and the distance between the hand and the target. A larger target can be reached faster than a smaller one at the same distance because the smaller one requires more targeting accuracy. Our 3D models had access to both of these parameters, so Fitts' Law became the inherent arm movement timing mechanism.

Further work with instructions had to wait for improved human models in *Jack*. Professor Mark Steedman joined the CIS department and we welcomed him into ANIML. Webber, Steedman, and I developed a roadmap for animation from instructions [4]. One of my Ph.D. students in ANIML, Jugal Kalita, studied the computational semantics of motion verbs [5]. Ph.D. students Brett Douville and Libby Levinson figured out how object shape and action requirements dictated hand shapes for grasping or manipulation [6]. Levinson completed her Ph.D. on this problem, calling it "object-specific reasoning." The underlying idea is simple but powerful: rather than having the virtual human compute how to grasp an object, the object itself *informs* the character how to approach and configure its hand. We can do that with computer animation, as both object and virtual humans share the same database. The problem is more difficult in robotics, where computer vision techniques or direct manipulation might be used to determine possibly unknown object shapes.

A good example of how object-specific reasoning works is opening a door by grasping and turning the knob and pulling the door toward oneself. To animate this, the doorknob tells the virtual human where it is in space, so it becomes a reach target. Fitts' Law computes the time needed to animate the reach. Once grasped, the doorknob axis tells the hand which way to rotate, and the arm movement is determined from the hand rotation. Finally, the door is made to rotate on its hinges and this now informs the hand that it is swinging inward. The arm moves because IK maintains the hand grasp on the doorknob. When the door opens sufficiently for passage, the hand grasp is released, and the arm can return to neutral or start its next task. These motions demand that the human model be compliant to external control, but with *Jack*'s interactive positioning facilities, these directives were easily executed.

Even basing movements on the desired result did not guarantee proper performance. Just because real people seem to accomplish simple tasks with little effort does not mean that they are equally easy for a digital model. It is overly optimistic to assume that computing body motions will be straightforward. The external impetus could cause or be restricted by self-collisions or external obstacles. Pose adjustments were essential to produce non-intersecting actions. Locomotor behaviors might even be necessary, especially if the target is not in the immediate reachable space of an arm or the body must step aside to pass through the doorway. We explored these implicit but essential motion strategies in several Ph.D. theses by Xinmin Zhao [7], Rama Bindiganavale [8], Tarek Alameldin [9], and Deepak Tolani [10]. Zhao populated *Jack* body segments with specific intersection check points to monitor point-to-point self-collisions and stop motions that would cause interpenetrations. Bindiganavale engaged torso motions to attempt collision avoidance. Alameldin explicitly computed a 3D model of the reachable space of a linkage so that it was easy to test whether or not a target was within reach or required additional pose or position adjustments. Tolani used optimal control methods to achieve overall collision-free motion paths.

Because even simple actions may require activating several lower-level human behaviors such as these grasps, steps, and postural adjustments, we started calling our approach "task-based animation." We would soon evolve a mechanism for organizing and executing such sequential and parallel activities as Becket's Parallel Transition Networks or "PaT-Nets" [11–13]. PaT-Nets were "finite-state machines"—stepwise software procedures with checkable transition conditions—that could watch for, and be triggered by, any environmental change or value. They could pass along parameters, such as a walk velocity, or watch for terminating conditions, such as satisfaction of a reach goal, that could then trigger further state changes. Multiple PaT-Nets could run effectively in parallel, allowing the control of human body systems (such as legs, limbs, and eyes) to be distributed across separate, manageable, programmable state machines. PaT-Nets could, as per the door example, break the human animation of opening a door into its necessary and semantically meaningful substeps. We would use PaT-Nets extensively for many projects over the next decade.

Although we began our animation from instructions work with NASA, we received a substantial boost from the aircraft maintenance community at the Wright-Patterson Air Force Base (WPAFB) Research Laboratory in Dayton, Ohio. The Air Force interest in computational human modeling had a long history, starting as early as 1972 with COMBIMAN, an aircraft pilot model led by Joe McDaniel [14]. Jill Easterly developed a successor system, Crew Chief, particularized to maintenance tasks with an extensive strength analysis capability [15].

In June 1990, Bonnie Webber and I spoke at a WPAFB "Workshop on Human-Centered Technology for Maintainability," a remarkable confluence of interested practitioners, human modeling experts, anthropometrists, software developers, and users [16]. Unlike the earlier NASA human factors meeting I had helped broker, the WPAFB workshop published proceedings and, as an especially valuable historical artifact, an *attendee list*. The list is probably as close to a "Who's Who" cross-section of human modeling efforts in 1990 as one will ever find. Our Air Force sponsors were there, of course, led by workshop organizers Ed Boyle, John Ianni, and Jill Easterly. Our NASA sponsors also attended: Barbara Woolford from JSC, Barry Smith from Ames, and Dr. Harold Frisch from Goddard Space Flight Center. Other *Jack* supporters in attendance from the Army included Brenda Thein of the Human Engineering Lab and Steve Paquette of Natick Labs. Several industry representatives in attendance would also be collaborators or *Jack* users in the coming years, including Dr. Richard Pew of BBN, Bruce Bradtmiller of Anthropology Research Project, Edgar Sanchez of Boeing, and Kevin Abshire of General Dynamics. Consultant John Roebuck—in my opinion, the reigning anthropometry expert of that time—would become a valued resource for us as well.

Ed Boyle, John Ianni, and Jill Easterly, our project contacts at WPAFB, were part of the Logistics Research Division of the Human Resources Directorate. Their mission was to ensure proper aircraft maintenance. This meant that the instruction manuals, called "Technical Orders" (TO), were comprehensive, correct, and up-to-date. The Air Force was in the process of converting printed TOs into electronic forms that would be easier to distribute, use, and update called Interactive Electronic Technical Manuals (IETMs). Our overall vision would be to "debug" IETMs by visualizing them with *Jack* maintainers and 3D aircraft models. Besides the Air Force sponsor, we collaborated with subject matter experts Kevin Abshire of Lockheed-Martin and Edgar Sanchez of Boeing. Soon we would see a role for simulation in the training space as well.

Immediately, we had some hefty problems to work on. Aircraft have thousands of parts, and not all of them are rigid. There are fasteners, piping, electrical wiring, flexible surfaces, moveable parts, and fluids. And hazards! We studied actual TO documents, watched tutorial videos, and visited the Air Force maintainer training facilities at Sheppard AFB in Wichita Falls, Texas. *Jack* was a commercial product by now, and we could concentrate on trying to get it to do some of the actions necessary for assembly, disassembly, and manipulation, such as part extraction and removal. In this vast problem space, we had to choose examples carefully. With Ph.D. students Ying Liu and Charles Erignac, we

Fig. 20.1 *Jack* interprets instructions to remove an aircraft power supply unit. It has mechanical, electrical, and hydraulic fluid connections. For safety, these devices must be accessed with tools or his hands, manipulated properly, and in a particular order. The green spheres indicate fluid connectors. For visual simplicity, the bulk of the surrounding aircraft model has been removed

examined the detailed steps needed to replace a power supply (Fig. 20.1). Action order, managing hydraulics, and maintainer safety were paramount considerations. Reaches into tight, constrained spaces required that Ph.D. students Ying Liu and Liwei Zhao revisit arm motion inverse kinematics with collision avoidance [17]. Reasoning about fluids was tricky, and Ph.D. student Charles Erignac built a qualitative reasoning engine for fluid motion. This constellation of problems attacked by human simulation resulted in an article in the flagship publication of the Association for Computing Machinery [18].

Our Air Force sponsors were serious about maintenance training. We connected with Patrick Vincent, who worked as a project liaison to Wright-Paterson through Litton/TASC (The Analytic Sciences Corporation) in Dayton. (We also shared an interest in dachshunds as pets.) For ten years, from 1998 through 2008, Vincent helped us navigate Air Force procurement processes to bring us a steady income stream for simulation developments. His preferred approach to funding involved creating a channel for funds to flow from the Human Resource Directorate into Litton/TASC, and then they would parcel out individual projects on an as-needed basis. This was a sweet deal since the entire project might be authorized for millions of dollars, but there was no obligation to spend it all, or even any of it, unless there were desirable subprojects. When a project arose that was deemed suitable for funds, a "Delivery Order" channeled the grant from the Air Force, through Litton/TASC, to us (or to others). These Delivery Orders behaved more like purchase orders than grants, but we were quite happy with their agility and customization qualities.

Once we had produced enough meaningful progress for our Air Force sponsors, mostly by controlling *Jack* through PaT-Nets, they decided that maintenance simulation needed to live under the umbrella of a real Air Force program. They called it DEPTH and issued a call for proposals. DEPTH covered the broader topic of human-centered design and offered the potential for significant funding to attract major defense contractor applicants.

DEPTH was too large and entailed other tasks for us to go it alone, so we partnered with another local Dayton contractor.

What ensued with our DEPTH proposal was a truly peculiar series of events. We generated our proposal and included the materials added by the local contractor. The proposal due date approached, and we dutifully had everything ready to go. Pushing proposals out through the University took a day or two (way back then), so we had it ready to FedEx out for timely arrival. The typical government proposal deadline is the close of business; I think in this case it was 4 pm Dayton time. That afternoon, I received a call from FedEx telling me that the proposal packet was undeliverable as the office it was addressed to was *closed*. The government rules are clear and strict: late by even a minute means you are out of contention. Since there was no one to receive the delivery, it was certainly late! What a disappointment after weeks of careful proposal preparation. Having been party to many unsuccessful proposals, I could take the bad news, but something didn't sit right. Soon enough, the situation became even more bizarre. I was told that the DEPTH project was awarded to General Dynamics. I hadn't even been aware that they were involved in human modeling and simulation. However, the proverbial lightbulb went on when I found out they had listed the University of Pennsylvania as a subcontractor! I guessed this was allowed. I wasn't an official co-Principal Investigator; we had submitted our own proposal. We weren't submitting two proposals simultaneously, either, which was forbidden. Thus, one way or another, we were part of the DEPTH team. While we did realize some research funding from DEPTH, overall we could not contribute at the level we ourselves had proposed. We re-directed the final batch of our DEPTH funding to a nearby small business owned by my former Ph.D. student, Dr. Paul Diefenbach, who completed the *Jack* C++ API and handled some loose ends in the interface code.

Among various Delivery Order funds, three especially stood out. The first, in 2007, involved evaluating wearable motion capture suits as a means of obtaining accurate human maintainer motions in complex environments. We bought one suit from Xsens and performed some accuracy assessments in our ReActor motion capture studio (which I'll describe later). We also took the suit to our local Navy base in Philadelphia, where an employee donned the suit and crawled around a shipborne air compressor unit. The advantage of the wearable suit was obvious here, as it ran on batteries and required no external sensing apparatus. We were disappointed at project completion because we had to send the suit back to the Air Force so they could use it. (NSF grants allowed the university to keep any purchased equipment.) The second Delivery Order provided us with an opportunity to study haptic feedback for real-time movement training. I will also explore this more later. The third Delivery Order involved using virtual reality (VR) as an actual task training aid.

We put some serious effort into this VR training project. It involved five Ph.D. students, some of whom were studying with other faculty. My core group consisted of Catherine Stocker, Ben Sunshine-Hill, and Joe Kider. Stocker was an undergraduate in computer science, then a Master's student, and we had already worked together on a project called

"Smart Events" [19]. This extended the object-triggering control idea, such as the door opening example described earlier, by recruiting virtual human agents from a pool of possibilities depending on their capabilities and availability. Joe Kider was my HMS lab manager at this time, so he was the default but capable project leader.

We took a couple of trips to the Air Force maintenance training site at Sheppard AFB. On our first trip, we scoped out a suitable task: securing and jacking up a fighter jet for other maintenance procedures. We felt this was something we could simulate, it had very specific task requirements, and it had distinct good and bad outcomes. For example, the jacks under the aircraft had to be positioned correctly and vertically, and then operated in turn to ensure that the plane remained level. Mistakes would be catastrophic if the plane tipped off the jacks. Our students built this simulation in PaT-Nets, and Stocker created a VR helmet interface so that a trainee could be immersed visually in the procedure. She also designed an experiment to test the teaching efficacy of the system. On our second trip to Sheppard AFB, we were allowed some time with an actual maintainer class for testing. We could not train everyone to use a VR display, so we evaluated three training modes: pre-existing video materials, our 3D virtual simulation shown on a 2D computer screen, and the original printed Technical Orders for the task. We allowed the maintenance students a short training session so that they would understand how to play the simulations and the videos. We found that participants did develop an increased awareness of hazards when training with simulations or videos, over just using the text materials [20]. While the results were encouraging, the Air Force appeared to lack the funding to pursue them further. The cost of VR displays and suitable high-performance computer workstations would have been prohibitive to further classroom adoption at that time. Although Stocker left the program before completing a Ph.D., she nonetheless gets credit for seeing this system undergo an actual training effectiveness evaluation.

References

1. N. Badler and J. Gangel. "Natural Language Input for Human Task Animation." Second Annual Workshop on Robotics and Expert Systems, Instrument Society of America, Houston, TX, June 1986, pp. 137–148.
2. J. Esakov, N. Badler and M. Jung. "An investigation of language input and performance timing for task animation." Graphics Interface '89, Morgan-Kaufmann, Palo Alto, CA, June 1989, pp. 86–93.
3. P. Fitts. "The information capacity of the human motor system in controlling the amplitude of movement." Journal of Experimental Psychology, Vol. 47 (6), 1954, pp. 381–391.
4. B. Webber, M. Steedman and N. Badler. "Narrated animation: A case for generation." ACL Workshop on Natural Language Generation, Dawson, PA, June 1990.
5. J. Kalita and N. Badler. "Semantic analysis of action verbs based on physical primitives." Cognitive Science Society 12th Annual Conference, Lawrence Erlbaum Associates, Hillsdale, NJ, 1990, pp. 412–419.

6. B. Douville, L. Levison and N. Badler. "Task Level object grasping for simulated agents." Presence 5(4), 1996, pp. 416–430.

7. X. Zhao, D. Tolani, B-J. Ting and N. Badler. "Simulating human movements using optimal control." Eurographics CAS '96, Seventh International Workshop on Computer Animation and Simulation, Futurescope, Poitiers, France, 1996.

8. R. Bindiganavale, J. Granieri, S. Wei, X. Zhao and N. Badler. "Posture interpolation with collision avoidance." Proc. Computer Animation, Geneva, Switz., 1994, pp. 13–20.

9. T. Alameldin, T. Sobh and N. Badler. "An adaptive and efficient system for computing the 3-D reachable workspace." IEEE International Conf. on Systems Engineering, Pittsburgh, PA, Aug. 1990, pp. 503–506.

10. D. Tolani and N. Badler. "Real time human arm inverse kinematics." Presence 5(4), 1996, pp. 393–401.

11. N. Badler, C. Phillips and B. Webber. *Simulating Humans: Computer Graphics, Animation, and Control.* Oxford Univ. Press, 1993. https://www.cis.upenn.edu/~badler/book/book.html

12. N. Badler, B. Webber, W. Becket, C. Geib, M. Moore, C. Pelachaud, B. Reich and M. Stone. "Planning and parallel transition networks: Animation's new frontiers." In Computer Graphics and Applications: Proc. Pacific Graphics '95, S. Y. Shin and T. L. Kunii (Eds.), World Scientific Publishing, River Edge, NJ, 1995, pp. 101–117.

13. J. Granieri, W. Becket, B. Reich, J. Crabtree and N. Badler. "Behavioral control for real-time simulated human agents." Proceedings of the 1995 symposium on Interactive 3D graphics (I3D), April 1995, pp. 173–180.

14. P. Krauskopf, J. Quinn, R. Berlin, W. Stump, L. Gibbons and J. McDaniel. "COMBIMAN Programs (COMputerized BIomechanical MAN-Model). Version 8 (User's Guide)," Defense Technical Information Center, ADA222735, Feb. 1989, https://apps.dtic.mil/sti/citations/ADA 222735 .

15. J. Easterly. "CREW CHIEF: A model of a maintenance technician." Defense Technical Information Center, ADA222669, May 1990. https://apps.dtic.mil/sti/tr/pdf/ADA222669.pdf

16. E. Boyle, J. Ianni, J. Easterly, S. Harper and M. Korna. "Human-Centered Technology for Maintainability: Workshop Proceedings." Interim Report ADA239090, Defense Technical Information Center, June 1991, https://apps.dtic.mil/sti/citations/ADA239090.

17. L. Zhao, Y. Liu and N. Badler. "Applying empirical data on upper torso movement to real-time collision-free reach tasks." SAE Digital Human Modeling Conference, Iowa City, IA, 2005 (published as paper 2005–01–2685, SAE Transactions Journal of Passenger Cars - Mechanical Systems).

18. N. Badler, C. Erignac and Y. Liu. "Virtual humans for automating maintenance task validation." Communications of the ACM 45(7), July 2002, pp. 56–63.

19. C. Stocker, L. Sun, P. Huang, Q. Wenhu, J. Allbeck and N. Badler. "Smart events and primed agents." Intelligent Virtual Agents (IVA) 2010.

20. C. Stocker, B. Sunshine-Hill, J. Drake, I. Perera, J. Kider and N. Badler. "CRAM it! A comparison of virtual, live-action and written training systems for preparing personnel to work in hazardous environments." Proc. IEEE Virtual Reality 2011.

Transitioning from Human Models to Virtual Beings

21

It would be disingenuous to claim that the *Jack* human models were the precursor to the realistic virtual beings of today. As one strand of the DNA, however, we did contribute on a number of interactive, structural, and algorithmic fronts. By the 1980s, many other researchers, corporations, and artists were making equally important advances and contributions to the virtual human DNA pool. Real-time motion sensing devices came on the market that allowed human motion capture. Artists vastly improved surface, face, hair, and clothing models. A wide range of artificial intelligence and cognitive science researchers were trying to add human sensing, thinking, reasoning, planning, gestures, and emotions to graphical body models.

The husband and wife team of Daniel Thalmann and Nadia Magnenat-Thalmann led one of the most active groups in virtual human technology. I met them when they were faculty at the University of Montreal in 1983. They invited me to be the plenary session speaker at their First Quebec Computer Graphics Symposium. Soon afterward, they moved back to their native Switzerland, where they took professorships at the École Polytechnique Fédérale de Lausanne and the University of Geneva, respectively. Over the next few decades, we would interact frequently. They kindly invited me to give tutorials at their institutions and sponsored other conferences, such as "Computer Animation," which I attended. They were a conference organizer power couple. Throughout their careers, they have remained committed to promoting computer graphics and animation through their research papers, conferences, community events, and journals. During my visits, I saw what directions they were working on and vice versa. I met many of their excellent Ph.D. students, including Prem Kalra, Ronan Boulic, Marcello Kallman, Hyewon Seo, Pascal Volino, and Soraia Musse. I would eventually collaborate with Musse many years later. I steered away from some of the Thalmanns' active areas, such as cloth modeling

N. Badler, *On Raising a Digital Human*, Synthesis Lectures on Computer Science, https://doi.org/10.1007/978-3-031-63945-6_21

and real-time VR, since there was always a vast research space to be explored in producing realistic humans. Their friendship was inspiring. Multiple visits to Europe were a cultural bonus.

New academically oriented conferences arose: Intelligent Virtual Agents (IVA) in 2001, Autonomous Agents and Multi-Agent Systems (AAMAS) in 2002, and the ACM/Eurographics Symposium on Computer Animation (SCA) also in 2002. In 2000, the human factors community began the Digital Human Modeling Conference (DHM), where I was honored to give the first Keynote talk. Artists and graphics practitioners, as well as researchers, excitedly reported their progress at the ACM SIGGRAPH Conference. I helped organize and teach well-attended human animation courses at the annual SIG-GRAPH Conference with the Thalmanns and others in 1991, 1992, and 1997 through 2000. We proselytized to hundreds of attendees. The creation of modern digital humans was an interdisciplinary, integrative, and evolutionary process. These emergent interconnections were publicly recognized in 1996–98 with three annual privately sponsored meetings in Los Angeles, simply called "Virtual Humans." The idea was surely to accelerate Hollywood's adoption of virtual humans, which had first appeared in *Westworld* in 1973. By 1997, *Titanic* exploited scores of virtual human digital doubles to populate the 3D ship model. Animator Shawna Olwen wowed the audience with her description of this process. However, they were only bodies following scripted directives, laboriously hand-crafted for one purpose and role.

One DNA thread from bodies to beings can be traced to a seminal paper led by Dr. Justine Cassell and presented at SIGGRAPH'94 [1]. This paper and video, two frames of which are shown (Fig. 21.1), demonstrated two *Jack* human models, George and Gilbert, conversing, gesturing, and speaking to each other about cashing a check at a bank. These two virtual people were *communicating*, not just executing task commands. Catherine Pelachaud created the animated faces. Professor Mark Steedman designed the dialog planner. Many Ph.D. students participated, ranging across AI, natural language understanding, speech acts, gesture generation, and graphics. Scott Prevost and Matthew Stone worked on the speech synthesis and language components, Brett Achorn built the gesture engine, and Brett Douville managed the graphics. Becket's Lisp API played a key role in linking these disparate components. Sitting atop the API, Becket's PaT-Nets controlled low-level agent behavioral motions through the *Jack* Toolkit.

This collaboration was fortuitous but accidental. Professor Bonnie Webber and I had been investigating relationships between movement and language since the early 1980s in our ANIML seminar. Webber's acquaintance, Dr. Justine Cassell, was teaching at Penn State (in State College, Pennsylvania), where she had joint appointments in Computer Science, Linguistics, and French. Webber invited Cassell to spend a semester visiting at the University of Pennsylvania. Cassell applied for and received a travel and research grant to do just that. Once at Penn, Cassell focused on her natural language collaboration with Webber. One day Cassell and I went to lunch. I told her about the human animation work we were doing with *Jack*. The idea mutually arose that perhaps we could exploit her

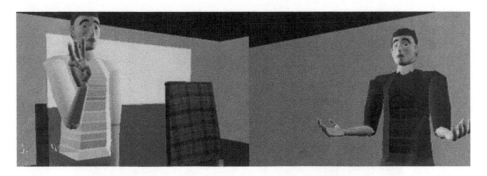

Fig. 21.1 George (left) and Gilbert (right) have an animated discussion about cashing a check at the bank

visit to try to build a "complete" talking, gesturing, and conversing digital human. Catherine Pelachaud was a postdoc at that time, and she led the project's implementation and realization. Thus began the "animated conversation" system shown at SIGGRAPH 1994. Two *Jack* characters, Gilbert and George, hold a conversation about cashing a check at a bank. The original SIGGRAPH paper has been reprinted several times. Cassell launched herself into a new career at the MIT Media Lab. We did not create the first animated "talking head," but we had demonstrated that many crucial language, planning, and graphics components could be integrated into a comprehensive digital human framework.

The animated conversation project pioneered the feasibility of controlling human figures through natural language. Cassell and Steedman pushed *Jack* toward autonomous behaviors in a constrained environment with movements limited to facial animation, arm, and hand gestures. We were not exploiting some of *Jack's* other movement features nor its real-time interface potential. In a subsequent attempt to perform language-driven activities, Master's student T.J. Smith and Ph.D. students Jianping Shi and John Granieri constructed a real-time animation system "*Jack*MOO" using lambdaMOO as the user interface front end [2, 3]. Inspired by online multi-user adventure-style text games, lambdaMOO was a Lisp-based system for parsing simple English commands, managing multiple users, and maintaining a dynamic, persistent, and shared textual data space.

Starting in 1995, under the aegis of an Advanced Research Project Agency grant through Philadelphia's Franklin Institute, we connected lambdaMOO to *Jack* human models via PaT-Nets and the *Jack* Toolkit. We modeled a decently rich 3D space resembling a rustic inn and added a number of behaviors that advantaged *Jack* features, such as climbing a ladder. By this time, Engineering Animation owned *Jack*, so we had free access to their extensive library of 3D models. We populated the inn accordingly, lending visual realism to the scenario (Fig. 21.2). Users could invoke digital human animations through simple English commands such as "sit down on a chair," "drink from your glass," "go to bed," and "leave the room." Some of these commands would invoke locomotive behaviors to navigate around the inn. These actions were often situationally ambiguous and

Fig. 21.2 The *Jack*Moo environment. The standing figure is the autonomous waiter who ensures everyone's drink is kept full. The other seated characters are the avatars of four simultaneous users. Control is achieved through simple sentence parsing and interpretation by lambdaMOO

required that *Jack* models make decisions and engage in simple planning, much like the animated conversation models. A notable distinction, however, was that these *Jack* models could be active *avatars* of one or more users. The avatar itself could make low-level decisions dependent on available resources and environmental constraints, such as choosing the closest unoccupied chair or navigating around obstacles. Actions could require a sequence of behaviors: commanding "go to bed" required that one's avatar get up from the table (if seated), navigate to the ladder leading upstairs, climb the ladder with an appropriate animation, navigate the hallway, and enter the bedroom. To enliven the space further, we included an autonomous (non-avatar) *Jack* "waiter." His job was to carry a pitcher and appear to pour water to fill any empty glass. If someone asked their avatar to "drink," when they were done and set the glass down on the table, the waiter would navigate over and refill their glass. If the waiter's pitcher were then empty, he would go to the kitchen to (invisibly) refill it. PaT-Nets were essential for executing these conditional activities.

JackMOO was restricted to the tasks it knew how to animate, such as sitting, climbing, drinking, and walking, and the decisions necessary to navigate the floorspace. Although interactive, *Jack*MOO was slower than true real-time, which could make participating in a session rather painful. The computers of the day were still a bottleneck for us. Other complications arose when Smith left the Ph.D. program, and Shi and Granieri went on to find alternative Ph.D. topics. Nonetheless, for us, *Jack*MOO was a huge step toward the realization of virtual beings driven by instructions in pseudo-natural language. Over the subsequent decade in the 2000s, we would take a deeper look into natural language instruction. We developed an additional rich and expressive representation for human actions that supported both language and graphical animation needs: PAR, the Parameterized Action Representation [4].

References

1. J. Cassell, C. Pelachaud, N. Badler, M. Steedman, B. Achorn, W. Becket, B. Douville, S. Prevost, and M. Stone. "Animated conversation: Rule-based generation of facial expression, gesture and spoken intonation for multiple conversational agents." Computer Graphics, July 1994, pp. 413–420.
2. T. Smith, J. Shi, J. Granieri, and N. Badler. "*Jack*MOO, An integration of *Jack* and lambdaMOO." Pacific Graphics, 1997.
3. J. Shi, T. Smith, J. Granieri and N. Badler. "Smart avatars in *Jack*MOO." Proc. Virtual Reality '99, Houston, TX, IEEE Computer Society, March 1999.
4. N. Badler, R. Bindiganavale, J. Allbeck, W. Schuler, L. Zhao and M. Palmer. "Parameterized Action Representation for virtual human agents." In *Embodied Conversational Agents*, J. Cassell, J. Sullivan, S. Prevost and E. Churchill (Eds.), MIT Press, 2000, pp. 256–284.

Locomotion and Gait

22

Bipedal walking is an essential human ability. We needed to have a walking capability so that *Jack* could navigate a 3D environment. We were not the first to study walking simulation. By 1978, psychologist James Cutting was already investigating the perception of gait using sparse, moving point lights at major body joints [1]. As our goals were directed toward locomotion synthesis on a whole-body model, several interesting problems arose. Walking is not just putting one foot in front of the other, finding angles of other joints in the legs, and playing repeat. We could implement that simple model easily, since we had inverse kinematics algorithms to find the knee and hip joint angles. Locomotion, however, is a complex of individualized and environmentally predicated behaviors. Walking—among an even wider range of possible locomotion gaits—is dependent on speed, body mass, skeletal structure, movement path, presence of obstacles, terrain characteristics, carried objects, personal gait style, and even emotional state. Extensive biomechanical research has been devoted to gait studies, both to understand the neuromuscular gait mechanism and to apply therapies to treat gait anomalies or perform physical rehabilitation.

From a computer animation perspective, gaits, such as walking or running, would appear to be straightforward cycles of joint rotations in the hip, knee, and ankle joints. In 1982, Ph.D. student David Zeltzer at the Ohio State University synthesized walking motions by embedding finite-state machines at leg joints to control the rotation angles during the walk cycle [2]. One of the first exercises expected of an artist's animated character is the production of a "gait cycle" loop that makes the character appear to locomote forever. Like other human animation groups, we needed to design a walking procedure to make our digital people go where they needed to go. Our approaches were not unique. We used human motion capture and programmed procedural joint motions to

© The Author(s), under exclusive license to Springer Nature Switzerland AG 2025
N. Badler, *On Raising a Digital Human*, Synthesis Lectures on Computer Science,
https://doi.org/10.1007/978-3-031-63945-6_22

animate walking. Ph.D. student Hyeong-Seok Ko created our first procedural attempt to model some of the complexities of walks [3]. He developed a realistic human locomotion controller based on the "zero-moment support point." This approach allowed gait and posture to change appropriately under the influence of external forces such as terrain contours, wind, or carried loads. Ko's controller could animate a *Jack* model walking in a windstorm carrying a dog and stepping over low obstacles. Motivation to manage movements under loads came from collaboration with Steve Paquette at the U.S. Army Natick Labs, which designed and evaluated soldier uniforms and equipment pack burdens. Ph.D. student Barry Reich also investigated terrain slope influences on locomotion [4]. Much later, we combined both to simulate gait on a ship at sea. Ph.D. student Seung-Joo Lee investigated stylistic differences in gait simulation caused by simple psychological factors such as joy or depression [5].

Throughout these gait studies in the 1990s, one visual aspect kept bothering me in our own work as well as in others. Our characters all walked, and their legs moved convincingly, but their *feet* did not look correct. The models always seemed to be "flat-footed": the joint rotation in the ankles was one-dimensional. I recalled reading the U.K. Ph.D. thesis on the knee joint and realized that there was a good biomechanical reason for the rotational complexity of the knee joint. The knee is not just a simple one-dimensional hinge joint, and neither is the ankle. In fact, the ankle and foot system is an entire complex of joints. It was no wonder that a simple planar rotation model of the leg joints, though easy to program, did not look correct to the observer's eye. By the time I realized that this was a significant problem, I was preoccupied with the *Jack* spin-off and other projects and had to put it aside for almost fifteen years until 2010.

Many approaches to gait simulation focus on maintaining balance. However, by forcing balance, the character seems to be "marching" or "stomping," pulled along by its feet rather than using them to control the body's center-of-mass (COM). In normal walking, the COM is moved *ahead* of the support polygon of the feet so that gravity pulls the torso forward and toward the unsupported side where the swing leg is off the ground. Normal gait cadence is fast enough to pull the swing foot forward and lock the knee to prepare for the force transfer at heel strike that lifts the pelvis on that side enough to stop the sideways fall and yet maintain forward momentum. A forward torso lean is essential for placing the COM ahead of the polygon formed by the foot supports. This is the reason that walking on a treadmill does not look the same as walking on solid ground.

Now consider the feet and especially the toes. We don't think about them when wearing shoes, but nonetheless, they still influence the gait pattern. The floor contact point moves from the heel (at heel strike) toward the toes (at toe-off). But this force transfer is nonlinear; otherwise, we would have toes that were all the same size and length. There's a reason we have big and little toes. The big toe is essential to supply more inward force at toe-off to counteract the sideways torso fall and bring the gait back toward the center-line. This requires a rather large force; hence, a large toe is needed to provide the extra force impulse for toe-off. The pressure pattern on the sole moves in a rough arc from the heel toward the little toe at the outer edge of the foot and then back toward the big toe, exiting along the toe direction line. In the 2010s, we verified this pattern with the sole pressure sensors installed in our lab. This force trajectory is further evidenced in the skeletal anatomy: the bones in the chain from the heel to the big toe have wide flat joint surfaces perpendicular to the line of force, a good design for a strong rolling motion, and a well-supported toe-off push. The shape of the foot sole also echoes the force trajectory: the outer edge is arc-shaped and fully supportive, while the inside (the instep) is actually shaped so that it normally makes no contact with the ground plane at all.

Now let's move up to the knee joint. There is some front-back sliding motion in the knee joint, as well as a "built-in" rotation: when the knee moves, it does not just bend (flex) or straighten (extend). There is also a slight rotational component in this motion. You can see this for yourself. Sit comfortably in a chair, feet flat on the floor. Now, lift a leg until it is straight in front of you. You will see your foot rotate slightly inward. Lowering your leg rotates it outward. You are "hardwired" to do this by your knee joint geometry.

What does this have to do with the walking pattern? First, notice that when the knee is at maximum extension, the joint is locked in the posterior (rear-ward) position. This locked position occurs at heel strike and transfers force from the lower leg efficiently through the upper leg to the hip, providing a nearly rigid leg chain. Sometime after support is established (near the position where the upper leg is vertical), the knee joint slides naturally to the anterior (forward), flexed position. Now, the knee rotation forces the foot to rotate outward at the toes—just what is needed to transfer the force to the inside big toe. As the support foot approaches toe-off, the knee is at maximum flexion and maximum anterior joint position. This effectively makes the upper leg slightly *longer* from the hip to the knee joint. This longer lever distance has a good side effect: the instantaneous knee joint is slightly further away from the hamstring muscle insertion points, so the hamstrings can pull the lower leg up with greater *speed* in this position. The advantage of this approach is that the trailing foot is quickly removed from the ground and lifted for clearance in the swing phase of the step. Most simulations have to use (artificial) curved paths for the toe trajectory (to some maximum clearance height). However, the toe path is more dependent on the knee joint motion: rapid up then straight, more than smooth up then down in an arc. This would also occur naturally from the contraction of the quadriceps that extends the lower leg forward and thus pulls the top of the lower leg backward at the knee joint.

I was personally disappointed not to further this line of gait simulation research. Ultimately, a good animator will eyeball the gait pattern so it looks correct, even if not physically and anatomically modeled. In another fortuitous turn of events, however, my colleague, CIS Professor Dimitris Metaxas, picked up this thread in the late 1990s. He furthered gait research through a collaboration with Physical Therapy Professor Jan Bruckner at the Thomas Jefferson University in Philadelphia. This is a great story in its own right and found its way into the popular press [6].

Once a character can walk, we want it to walk somewhere and not usually run into things. Navigation is another popular area in agent animation. Navigation means reaching a goal, skirting obstacles, and avoiding collisions with other human agents. For an individual, navigation mostly reduces to determining and following a straight or curved walking path on the ground. Alternatively, explicit footstep positions can indicate the gait path, similar to dance step instructions. Adding an artificial vision sense to the human is a handy way to make it appear to relate to the upcoming environment and thus anticipate and correct for the presence of obstacles [7]. As in all human behavior, real-life situations can be complex, requiring more advanced reasoning and decision-making. The simultaneous application of navigation procedures to a group of people is a fundamental requirement in crowd simulation. We'll return to this topic later.

References

1. J. Cutting. "A program to generate synthetic walkers as dynamic point-light displays." Behaviour Research Methods & Instrumentation, 10, 1978, pp. 91-94.
2. D. Zeltzer. "Motor control techniques for figure animation." IEEE Computer Graphics and Applications 2(9), November 1982, pp. 53-59.
3. H. Ko and N. Badler. "Animating human locomotion in real-time using inverse dynamics, balance and comfort control." IEEE Computer Graphics and Applications, 16(2), March 1996, pp. 50-59.
4. B. Reich, H. Ko, W. Becket and N. Badler. "Terrain reasoning for human locomotion." Computer Animation '94, Geneva, Switzerland, IEEE Computer Society Press, Los Alamitos, CA, 1994, pp. 76–82.
5. K. Ashida, S.-J. Lee, J. Allbeck, H. Sun, N. Badler, and D. Metaxas. "Pedestrians: Creating agent behaviors through statistical analysis of observation data." Proc. Computer Animation November 2001, Seoul, South Korea, pp. 84–92.
6. W. Hively. "Bruckner's Anatomy." *Discover* online, 1999. https://www.discovermagazine.com/technology/bruckners-anatomy. Accessed February 12, 2024.
7. O. Renault, N. Magnenat-Thalmann and D. Thalmann. "A vision-based approach to behavioral animation." Visualization and Computer Animation, Vol. 1(1), 1990, pp. 18-21.

A Virtual Human Presenter

23

By the late 1990s, some digital human animators began to adopt *mark-up languages* as a simpler alternative to on-the-fly language understanding. These mark-ups were "meta" annotations added to pre-existing natural language text. Mark-ups communicated various verbal and nonverbal behaviors to the character—such as text stress, arm gestures, posture, emotional displays, and eye gaze—directly synced with the words to be spoken [1]. Our success with Gesture *Jack* had landed Justine Cassell a faculty position at the MIT Media Lab, where she continued to explore this paradigm with her students [2].

An excellent use case for animated humans working from predefined text procedures was *pedagogical* agents: digital instructors who could engage a learner, especially in spatialized tasks. Because *Jack* was embeddable in other software systems, he was adopted, modified (by removing his legs), and renamed *Steve* by James Rickel and Lewis Johnson at USC. Their *Steve* agent could guide a learner through a complex 3D mechanical maintenance procedure and, because it used an embedded AI planner, could recover from errors and explain its actions [3].

A resident visitor to our lab at Penn, Professor Tsukasa Noma of the Kyushu Institute of Technology in Japan, used a mark-up scheme to implement a "*Jack* presenter." Dr. Noma observed that *Jack's* abilities aligned well with the behaviors expected of a TV weatherperson: speaking from a pre-existing script, pointing to locations on a large projected map backdrop, looking at a virtual camera, and walking back-and-forth on a stage to best position itself for the pointed map references. Noma and my Ph.D. student Liwei Zhao created the "*Jack* Presenter" system (Fig. 23.1) using mark-up annotations embedded in text [4]. For example, suppose this is the desired spoken script:

N. Badler, *On Raising a Digital Human*, Synthesis Lectures on Computer Science, https://doi.org/10.1007/978-3-031-63945-6_23

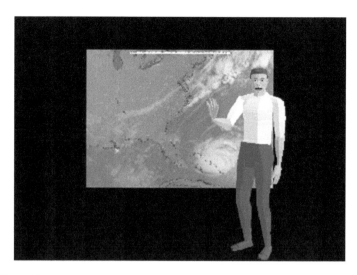

Fig. 23.1   The *Jack* presenter "weatherman."

> *The application area of the presenter is potentially so vast. For example, I can be a weather reporter. Hurricane Bertha is now to the east of Florida peninsula. It is now going north. New York and Philadelphia may be hit directly. Take care.*

This is manually marked-up with explicit control signals:

> *The application area of the presenter is potentially so vast. For example, I can be a weather reporter.* \board{berthapanel.gif, berthapanel.vbm} \point_idxf{bertha} *Hurricane Bertha is now to the east of* \point_back{florida} *Florida peninsula. It is now going* \point_move{bertha, north} *north.* \point_idxf{newyork, philadelphia} *New York and Philadelphia may be hit directly.* \gest_warn *Take care.*

Coordinating all the body joint movements from this high-level mark-up is nontrivial since motion competitions and conflicts may arise from the mark-up directives. Such clashes were mitigated by a hierarchy of Pat-Nets controlling groups of body parts, such as a walk-net, an arm-net, a hand-net, a head-net, and a speech-net. Although the *Jack* weatherman was not as "smart" nor interactive as Rickel and Johnson's *Steve* agent, our *Jack* presenter had legs, locomotion, and collision avoidance capabilities. Locomotion issues are tricky, and others often avoided them entirely to focus on head, speech, and arm movements. Even Meta's much-touted avatar release of 2022 did not incorporate legs until later in that year [5]. We did not appreciate in 2000 that virtual human presenters would reach a degree of realism sufficient to replace some real announcers by the early 2020s.

References

1. K. Thorisson, H. Vilhjalmsson, C. Pelachaud, S. Kopp, N. Badler, W. L. Johnson, S. Marcella and B. Krenn. "Representations for multimodal generation: A workshop report." AI Magazine 27(1), Spring 2006, p. 108.
2. J. Cassell, H. Vilhjálmsson and T. Bickmore. "BEAT: the Behavior Expression Animation Toolkit" SIGGRAPH '01: Proceedings of the 28th annual conference on Computer Graphics and Interactive Techniques, August 2001, pp. 477–486.
3. J. Rickel and W. Johnson. "Integrating pedagogical capabilities in a virtual environment agent." Proc. of the First Int'l Conf. On Autonomous Agents, 1997, pp. 30–38.
4. T. Noma, L. Zhao and N. Badler. "Design of a virtual human presenter." IEEE Computer Graphics and Applications 20(4), July/August 2000, pp. 79–85.
5. K. Holt. https://www.engadget.com/meta-avatars-metaverse-legs-vr-virtual-reality-facebook-instagram-whatsapp-190937062.html/ . Accessed August 26, 2023.

Movement Notations

24

Embarking on a Penn faculty research career in the late 1970s, I still believed in the notion that computers ought to be able to see and describe moving scenes. Although clearly a crucial component of what we talk about, human motion description was missing from my thesis. My animation system made it easy to animate rigid objects such as cars and trains, but it did not have the nested rotation capability to do an articulated human. It was also limited to lines and polygonal surfaces. Finally, I lacked the time to explore human animation in any depth due to my thesis deadline.

I discovered that another young Penn CIS faculty member, Professor Steve Smoliar, had an interest in dance, especially ballet. I surmised that the dance community had already developed good schemes for describing human motion, and it would be worthwhile to explore this direction with him. Fortuitously, Smoliar's partner at that time, Lynne Weber, had been a ballet dancer. She enrolled at Penn for a computer science Master's degree in Engineering and an MBA at the Wharton School. Weber provided key links into the world of dance through her close association with the Dance Notation Bureau in New York.

Up to this time, I had no overt interest—or skill—in dance. I don't recall even attending a dance performance other than a guest Master Class given by Jacques D'Amboise while I was an undergraduate in the College of Creative Studies at UCSB. Perhaps my new dance interest was an echo of my ninth-grade crush on the smartest girl in the class, Anita Paciotti. We never progressed beyond a math rivalry, and at the end of ninth grade, I moved away to a different school district. Paciotti later joined the San Francisco Ballet Company and became their Principal Character Dancer and then Ballet Master.

Smoliar, Weber, and I began a deep dive into dance representations that might be compatible with digital structures [1]. We dismissed the Benesh and Eshkol-Wachman systems

© The Author(s), under exclusive license to Springer Nature Switzerland AG 2025
N. Badler, *On Raising a Digital Human*, Synthesis Lectures on Computer Science,
https://doi.org/10.1007/978-3-031-63945-6_24

from consideration because their sequential rotation bases made them prone to individual deviations, cumulative misinterpretation, and errors. We therefore mostly explored Labanotation [2] and built a full syntactic description of its structure and symbology [3]. Labanotation is a graphic representation of instructions to move body parts. Some of the key notions in Labanotation included reach goal positions, arbitrary locomotion patterns, motion paths on the floor, and body part contacts. Timing was explicitly graphed on a vertical staff. Multiple individual performances could be easily aligned in time and space. Master's students Maxine Brown and Vicki Hirsch built graphical editors for Labanotation staff symbology and floor plans, respectively, which were exemplary interactive systems at that time. LifeForms [4], an extensive project at Simon Fraser University led by Professor Tom Calvert, succeeded in driving human stick figures through a Labanotation interface, but the resulting movements were perhaps unexpectedly generic and robotic. This observation led us to explore alternatives.

The breadth of descriptive mechanisms in Labanotation pushed us to investigate what new algorithmic capabilities would be needed if our digital representations were to be used for realistic graphic animations [5]. We identified five specific problems: scheduling movements that occur concurrently, computing the motion of three-link chains (arms) with inverse kinematics, processing contacts, moving the center of gravity and maintaining balance, and adjusting limb twists for a standard orientation. These became topics for several subsequent Ph.D. theses and papers during the 1980s and 1990s. Among the most notable were studies by Ph.D. students James Korein, Deepak Tolani, and Jianmin Zhao on reach inverse kinematics and Hyeong-Seok Ko's locomotion model [6].

Impressed with the world of human movement notation, in 1982, I took advantage of attending a conference in London, England, to avail myself of an opportunity to visit the reigning "guru" of Labanotation, Ann Hutchinson-Guest. We had mutual connections through Lynne Weber and others at the Dance Notation Bureau. This would be my third "pilgrimage": this one to visit the legendary Hutchinson-Guest. She invited me to meet at her home, and we had a cordial discussion. I don't think she had ever considered how computers might relate to the world of dance. In 1982, they certainly weren't ready to replace dancers.

Labanotation is a kinematic mechanism for describing human motion: indicating *what* to do but not necessarily *how* to do it. To copyright a dance in the US, a notated score is required. A movie or film of a dance performance is not sufficient. The notated score left unspecified the artist's actual nuanced interpretation during the performance, exactly analogous to the case with music notation. Thus, while movement timings were explicit in Labanotation, how the movement was performed during that time was not. Not surprisingly, the inventor of Labanotation, Rudolph Laban, also realized this limitation. With colleagues, he attempted to define an additional notation to characterize what *qualities* a performed motion appears to project to an observer. He called this extended system Laban Movement Analysis (LMA) or Effort-Shape Notation [7]. LMA opened a route into the human movement space that was lacking in purely kinematic approaches and seemed imminently compatible with my desire to describe human motion in language terms.

While my students were occupied with the TEMPUS system and other Ph.D. topics, my own background task was to determine how LMA might be turned into an explicit representation for controlling a human animation. The LMA Effort-Shape parameters were tantalizingly named, but in themselves did not provide any clear paths to kinematic realization. The Effort dimensions were Space, Weight, Time, and Flow; Shape dimensions were Enclosing-Spreading, Rising-Sinking, and Retreating-Advancing. Each dimension had extremes characterized by language terms. I tried to design a workable LMA implementation framework, but it had gaps I could not yet address [8]. I speculated that a physics-based simulation approach might be necessary but did not know where the physical parameters would come from or how individual body size and shape issues should be accommodated.

My vision was finally realized a decade later when my Ph.D. student Diane Chi and visiting Ph.D. student Monica Costa (from Pontifícia Universidade Católica do Rio de Janeiro) created the breakthrough EMOTE system presented at SIGGRAPH 2000 [9]. We now had a human figure whose motion could originate from any source, such as motion capture or key-pose animation, and the perceptual *qualities* of that motion could be varied by explicit Effort-Shape dimension "dials." Chi's realization implemented LMA without resorting to any virtual physics computations. The LMA dials controlled approximately three dozen lower-level animation parameters. Many of these were themselves inspired by the hand-drawn animation guidelines prevalent in classic animated cartoons and movies [10].

The marriage of EMOTE and *Jack* spawned numerous offspring. Ph.D. student Liwei Zhao applied EMOTE to gestural motions. He used EMOTE to change the apparent personality of an animated speaker or presenter [11]. This application turned out to be a precursor to our later work with a more principled approach to personality manifestation. Zhao also demonstrated how EMOTE could implement certain adverbial expressions, changing the apparent force of a motion without invoking physics. This adverbial work greatly influenced a project led by Professor Mitch Marcus' Ph.D. student Matt Heunerfauth to animate American Sign Language (ASL) [12]. ASL is a true natural language and has expressive capabilities beyond mere sign choice. Actual hand and arm motions may directly imitate or generalize movement verbs. Adverbs further modify these motion performances, often in a physically sympathetic manner. These modifiers are called "classifier predicates" in ASL. EMOTE was a valuable tool for covering more of the possible space of ASL utterances as gestural motions because the Effort dimensions realized classifier predicates for adverbs and the Shape dimensions enabled classifier predicates for certain spatial and size adjectives.

Ph.D. students Koji Ashida, Seung-Joo Lee, and Harold Sun applied EMOTE parameters to locomotion cycles [13], yielding depressed, energetic, or even asymmetric gaits, such as when carrying loads. Following on that success, Ph.D. student Meeran Byun explored mapping EMOTE parameters onto a human face model [14]. Although the results were interesting, there was no empirical evidence that LMA parameters would apply to faces with the same perceptual connotations. This remained a latent curiosity, and we abandoned further work on this approach when Byun had to withdraw from the Ph.D. program for personal reasons.

EMOTE did have a tour of duty beyond our lab. The Institute for Creative Technologies (ICT) at the University of Southern California launched in 1999 as an Army-sponsored research center with an overall mission to bring together "Hollywood"-style computer graphics and emergent AI techniques to control digital humans. I was appointed to their Technical Advisory Committee and watched their annual progress at on-site meetings. Dr. Paul Debevec developed his light-stage technologies at ICT. This soon became the de facto technique for obtaining 3D model surface reflectance functions for real actors who could be virtually relit and thus rendered by arbitrary ambient lighting environments. My small role here was to advocate annually for Debevec's team. Almost everyone else in ICT had an AI focus. Researchers at ICT were authoring interactive, graphical, human-centric training systems for the Army. To portray people and soldiers, they used DI-Guy, a set of realistic digital human models built and distributed by Boston Dynamics. To enhance expressiveness, ICT adapted EMOTE to control DI-Guy animations. My relationship to DI-Guy is another story, which I will relate later.

I felt in a vague funk in the early 2000s. *Jack* had moved on, my funding was generally waning, and my collegial landscape was in flux. ICT looked exciting, well-funded, and centered on topics I had already found intellectually stimulating. Their initial Director came from the "Hollywood" side, and it appeared that they now wanted a technologist to fill that role. My background in graphics, artificial intelligence, and digital humans seemed to fit perfectly. They asked me if I would like to interview for the Directorship. Clearly flattered, I agreed to an on-site interview at USC, since a computer science faculty appointment would have been part of the deal. A short while after that visit, I received a call from the Army, asking me to come for a second interview. Since it was Army money, they had a say. Now that this was sounding serious, I became nervous and asked them to delay the interview while I thought it over. During that time, I attended another meeting to discuss a forthcoming Army research project. The Director of the Robotics Institute at Carnegie Mellon University, Professor Matthew Mason, was also in the audience. I knew him from his robotics research, so I felt comfortable asking him about what it was like to lead a large, well-funded, internationally acclaimed research enterprise. His reply was crisp and informative: in essence, such a role is about promoting other people's work rather than one's own. This sealed my decision, which had been turning negative already, as I was diagnosed with cancer and did not want to uproot my life to move to a new situation in southern California. I declined the Army interview and withdrew from the search.

References

1. N. Badler and S. Smoliar. "Digital representations of human movement." ACM Computing Surveys 11(1), March 1979, pp. 19-38.
2. A. Hutchinson Guest. *Labanotation: Or, Kinetography Laban: the System of Analyzing and Recording Movement*. Taylor & Francis, 1977.
3. L. Weber, S. Smoliar and N. Badler. "An architecture for the simulation of human movement." Proc. ACM National Conf., Washington, DC, Dec. 1978, pp. 737–745.
4. T. Calvert, J. Chapman and A. Patla. "Aspects of the kinematic simulation of human movement." IEEE Computer Graphics and Applications 2, 1982, pp. 41-50.
5. N. Badler, J. O'Rourke and B. Kaufman. "Special problems in human movement simulation." ACM SIGGRAPH Computer Graphics 14(3), Summer 1980, pp. 189–197.
6. H. Ko and N. Badler. "Animating human locomotion in real-time using inverse dynamics, balance and comfort control." IEEE Computer Graphics and Applications, 16(2), March 1996, pp. 50-59.
7. I. Bartenieff and D. Lewis. *Body Movement; Coping with the Environment*. New York: Gordon and Breach, 1980.
8. N Badler. "A representation for natural human movement," in *Dance Technology* I, J. Gray (Ed.), AAHPERD Publications, Reston, VA, 1989, pp. 23-44.
9. D. Chi, M. Costa, L. Zhao and N. Badler. "The EMOTE model for Effort and Shape." ACM SIGGRAPH '00, New Orleans, LA, July, 2000, pp. 173–182.
10. J. Lassiter. "Principles of traditional animation applied to 3D computer animation." ACM SIGGRAPH Computer Graphics, 21(4), July 1987 pp. 35-44.
11. L. Zhao, M. Costa and N. Badler. "Interpreting movement manner." Proc. Computer Animation Conf., IEEE Computer Society, Philadelphia, PA, May, 2000.
12. L. Zhao, K. Kipper, W. Schuler, C. Vogler, N. Badler and M. Palmer. "A machine translation system from English to American Sign Language." Proc. Association for Machine Translation in the Americas, 2000.
13. K. Ashida, S.-J. Lee, J. Allbeck, H. Sun, N. Badler, and D. Metaxas. "Pedestrians: Creating agent behaviors through statistical analysis of observation data." Proc. Computer Animation November 2001, Seoul, South Korea, pp. 84–92.
14. M. Byun and N. Badler. "FacEMOTE: Qualitative parametric modifiers for facial animations." Symposium on Computer Animation 2002, San Antonio, TX, July 2002.

Motion Capture

25

Many tools we take for granted today were nonexistent in the 1980s. Although Ivan Sutherland had jump-started interactive computer graphics two decades prior with his 1963 Ph.D. thesis [1], CAD design software systems ran on large mainframe computers with vector displays and awkward user input devices such as "light pens." In 1984, Wavefront Technologies offered one of the first viable commercial 3D computer graphics modeling systems. In fact, as their Wikipedia page states, "… there were no off-the-shelf computer animation tools available at the time" [2]. Starting with O'Rourke in the late 1970s, we had to code our own 3D modeling tools. My TEMPUS team had to build their own 3D model input tools, but we never expressed any pretensions that TEMPUS was a CAD system on its own.

Somehow, probably through a SIGGRAPH exhibit, I found out about a new 6-axis digitizing system developed by Polhemus and offered around 1981 [3]. I purchased one for the lab, and I think we received a unit with serial number 4. It was a two-foot square desk on a four-foot-high cabinet and included a Vermont maple stool so the operator could sit while working (Fig. 25.1). The business end of the digitizer was a pen-like stylus and a cubical housing for electrical coils. Inside the cabinet sat a large electromagnet. The numerical coordinates of the six degrees of freedom (three rotations and three translations in space) of the stylus tip were computed from the magnetic eddy currents generated in the stylus coils. The advertised purpose of this device was to digitize points in 3D: one placed a physical object on the table and touched surface points that were recorded and transmitted to the host computer when a foot pedal was depressed. We quickly learned that touching the object moved it, so the object had to be taped down during the data acquisition process.

© The Author(s), under exclusive license to Springer Nature Switzerland AG 2025
N. Badler, *On Raising a Digital Human*, Synthesis Lectures on Computer Science,
https://doi.org/10.1007/978-3-031-63945-6_25

Fig. 25.1 Ginny Badler uses
the Polhemus stylus to digitize
3D points on a mug

One of my first uses of the Polhemus was to digitize pottery sherd profiles and cross-sections. Ginny's archaeology work required that she measure and draw hundreds of sherds. I volunteered to turn these sherds into publishable drawings. I spent literally months manipulating B-spline curves in CorelDraw to produce acceptable pottery drawings. I mention this mostly because the Polhemus wasn't particularly good for this, and we ended up doing all the sherd data collection by manual means anyway. We had this interesting device in the lab just sitting around.

Something I came to appreciate as a faculty member, and is not so well understood by the public, is that research is not always dictated top-down in a lab. Having interesting "stuff" lying around is bait that attracts the right curious student. The unused Polhemus caught the attention of one of our computer science undergraduates, David Baraff. I had good reason to encourage him. When Baraff applied to Penn, he included a 3/4" U-Matic videocassette with his materials. It was common for creative students to include a portfolio, but the Admissions Office had no idea what to do with this video tape. Someone in Admissions heard that I had a video suite and sent the tape over to me for review. It was an amazing animation of 3D textured objects. Baraff apparently had access to his father's computer at Bell Labs and produced the modeling, animation, and rendering on his own.

Baraff started using the Polhemus, but not as a digitizer! Rather, he explored using it as a generalized *tool* whose motions could be recognized as those of a specific nature. Thus, the probe motion could emulate the motions of using a hammer, saw, or screwdriver. Even more significantly, these were motions of human activities. Baraff had shown us that the Polhemus 6-degree-of-freedom inputs could capture human arm *motions*. Although written up in 1986 as a CIS Technical Report aptly entitled "Handwaving in computer graphics: Efficient methods for interactive input using a six-axis digitizer," it was never published. It was alluded to in our subsequent work [4] but remains a missed opportunity.

Instead of further pursuing that particular direction, Baraff began working with Master's students Kamran Manoochehri and Graham Walters on interactive posing of the TEMPUS "triangle man" articulated human body on a Silicon Graphics workstation. They developed an iterative real-time algorithm that continuously re-posed the figure to best satisfy 3D hand and foot reach constraints directly imported from Polhemus inputs. Body integrity was preserved. The figure would do its best to satisfy the required reach points simultaneously without breaking apart, just like a physical mannequin [4].

Baraff went on to Cornell University for his Ph.D. under Professor Don Greenberg. There he laid the foundations for physics-based modeling, especially of clothing. After his Ph.D., Baraff joined Pixar, where he led much of their animation research. In 2006, Baraff, along with Pixar colleagues Michael Kass and Andy Witkin, were honored with a Motion Picture Academy Scientific and Technical Achievement Award for cloth simulation.

References

1. I. Sutherland. "Sketchpad: A Man-Machine Graphical Communication System." Ph.D., Massachusetts Institute of Technology, 1963.
2. https://en.wikipedia.org/wiki/Wavefront_Technologies.
3. https://encyclopedia2.thefreedictionary.com/digitizer.
4. N. Badler, K. Manoochehri and G. Walters. "Articulated figure positioning by multiple constraints." IEEE Computer Graphics and Applications, June 1987, pp. 28–38.

With our appetites whetted by the possibilities of real-time 3D input, we next acquired another new motion sensing system, this time from Ascension Technologies. This device externalized the electromagnet and allowed multiple cube-like sensors to be placed any-where within a six-foot radius of the magnet. In other words, the sensors could be strapped onto body parts! The active space was restricted to the magnetic field, and the sensors, though small, were tethered to a connector port. By attaching just six sensors to wrists, ankles, head, and waist, we could now use actual human movements and our inverse kine-matic code to drive the full graphical *Jack* human model [1]. A user could move their 3D *avatar* through body motions.

We now possessed all the tools for a VR experience that included a model of self. We liked working with Ascension, and they had come out with a room-sized motion capture system called the Reactor. We received an NSF Equipment Grant in 2002 to purchase a Reactor and additional equipment to turn one side of the Reactor frame into a rear-projected stereo display wall. We installed this system in one of the former PC classrooms in our lab space (Fig. 26.1). We then had a one-sided display room in which a user could experience a 3D stereo environment and simultaneously control their avatar and camera view position.

The Reactor itself was a cubical frame with photocells spaced closely along each of its eight edges. The subject wore a special body suit wired with miniature LEDs, a battery pack, and a wireless communication module (Fig. 26.2). By rapidly pulsing through the LED set, the photocells could detect their illumination and triangulate their 3D spatial positions in real time. Approximately 20 LEDs could be tracked at 30 frames/second. If one wished, they could even be divided among two individuals in the capture space. In

© The Author(s), under exclusive license to Springer Nature Switzerland AG 2025   119
N. Badler, *On Raising a Digital Human*, Synthesis Lectures on Computer Science,
https://doi.org/10.1007/978-3-031-63945-6_26

Fig. 26.1 A 3D model of the LiveActor motion capture space. The foreground wall (in black) was the stereo backprojection surface

either single or dual-person use, LED occlusions could be a problem, but having a whole-body model meant we could often interpolate missing sensors or resort to using IK for elbows and knees.

We had to do some construction work to ensure that the Reactor system was installed correctly. The interior of the cubical frame had to be raised off the floor about six inches. One of my lab IT employees, Matt Beitler, volunteered to do the carpentry. In what was likely a violation of Penn's labor and construction agreements, I delivered the plywood to the lab, and Beitler constructed and painted the platform. We were ready for the Reactor installation. To drive the display wall, we contracted with Eon Reality. They supplied stereo rendering software and a data-flow programming language interface. The stereo projectors were touchy but were adequately installed and managed by a subcontractor, Professor Karim Abdel-Malek of the University of Iowa. Abdel-Malek would later build his own human modeling system, called *Santos*, as a *Jack* competitor [2].

We named this facility "LiveActor" [3]. For the LiveActor inauguration, we invited our local U.S. Congressional Representative, Curt Weldon. The idea was to utilize this setup as a potential model for real-time VR training for the military. We did some press and photo-ops. Unfortunately, Weldon was soon embroiled in a scandal and was never able to help us directly in this quest.

We were still interested in human input for VR and started to look at other sensing modalities. One of my Ph.D. students, Aaron Bloomfield, began studying haptics (touch) and vibrotactile sensing [4]. A "tactor" is a small motor with an off-center weight. When energized, it vibrates. Positioned against the skin, it creates a tangible sensation.

Fig. 26.2 The Reactor motion capture system. Jeffry Nimeroff is wearing the body suit, battery pack, transmitter/synchronizer, and LEDs. He is controlling an "enhanced" Jill model. Many observers remarked that something looked vaguely "off" in Jill's animation: it was because she was mirroring Nimeroff's motion, not a woman performer

I had heard about a tactor suit designed for jet pilots that would allow them to sense the direction of gravity ("down") even during high-g maneuvers. Bloomfield designed and constructed his own wearable tactor sleeve with 16 cell phone vibration motors [5] (Fig. 26.3).

Bloomfield's motivation was to allow a maintenance worker watching immersive VR visuals of some mechanical assembly to sense when their arm contacted a solid component. Tactile feedback and physical forces are not the same. The imaginary Star Trek Holodeck somehow creates real forces on its participants (it is science fiction, after all!), for example, for hand-to-hand combat training. We had to be content with just the sense of touch. Such vibrotactile sensations would then hypothetically motivate the wearer to execute small corrective arm movements to eliminate the vibrations and, hence, the virtual collisions. While this was a sound assumption if one tactor was activated, it was not possible to reliably predict a corrective movement direction for some arbitrary set of activations. The direction of compensatory motion could not be ascertained without coordinated visual feedback. By placing the subject wearing the sleeve inside the Reactor, the spatial locations of the tactors on the arm could be detected and used to drive the virtual model arm in the 3D maintenance space. Bloomfield was my only Ph.D. student who built an integrated hardware and software system. He was also the most earnest LiveActor user.

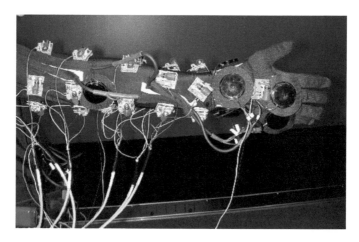

Fig. 26.3 Bloomfield's arm sleeve with 16 tactors providing localized vibratory stimuli

After his Ph.D. and a brief postdoc with me, he took a faculty position at the University of Virginia, where he garnered numerous teaching awards.

Another Ph.D. student, Liwei Zhao, exploited the Reactor's whole-body motion capture capability [6]. Zhao revisited my old desire to have computers describe human movement in language. That was still too big a step, so instead he used 3D motion capture data from the Reactor and output the LMA qualities of the movement. We recruited two dancers through the Dance Notation Bureau as consultants. They were also LMA notators. Zhao captured them producing various movements across the LMA descriptive spectrum. Since Zhao had ground truth from the dancers' own knowledge and performances, these motion capture datasets were used to train neural networks to recognize the LMA qualities. It made a pretty nifty demo. The motion capture subject could perform rather arbitrary whole-body motions, and the LMA qualities were computed and displayed on the LiveActor screen in real time.

Beyond Bloomfield's tactors and Zhao's LMA detector, few other insights or papers arose from LiveActor. I was ultimately disappointed that the whole concept as a training aid never quite took off. A retrospective postmortem yields some insights. The notion of display room VR was not new, having been popularized and named a CAVE® by Carolina Cruz-Neira and colleagues years before [7]. Our premise was that whole-body input would add a human dimension beyond immersive visuals. *Jack* even appeared as the only human model in the CAVE's debut demonstration at SIGGRAPH 1992 [8]. However, no Ph.D. student took ownership of LiveActor enough to do the user interface science that such a facility could have enabled. This was perhaps the fatal flaw. I should have known better. My own guidance for projects had always been "people first, hardware second." Although we did make some use of LiveActor, it violated this adage, leading to an unexpected level of neglect.

I was under significant personal and medical stress during the mid-2000s after being diagnosed with bladder and prostate cancers. I could not devote the necessary time and energy needed to exploit the LiveActor concept. One of the steroid medications I was taking caused unexpected memory loss. I still call 2003–2005 my "dark ages." My publication record from that period appropriately reflects a paucity of progress. Were it not for the self-motivated Ph.D. students I had at the time, such as Bloomfield, Zhao, and especially my lab manager and Ph.D. student Jan Allbeck, my research enterprise would have been even more dismal. Allbeck became my principal co-author, though she did virtually all the work. She even stepped in to travel and deliver a Keynote Talk I had agreed to do, but could not attend.

At some human factors conference luncheon, I was sitting next to a faculty member from the University of Denver. He was lamenting that they could not afford a motion capture system for their engineering students, many of whom were from disadvantaged backgrounds. I asked him if perhaps he would like to adopt our Reactor system. It was already decommissioned and just occupying space. "Sure!" he said. We soon disassembled the entire system and shipped it to the University of Denver. At least through 2010, they were still actively using it [9].

References

1. N. Badler, M. Hollick, and J. Granieri. "Real-time control of a virtual human using minimal sensors." Presence 2(1), 1993, pp. 82-86.
2. K. Abdel-Malek, J. Arora, J. Yang, T. Marler, S. Beck, C. Swan, L. Frey-Law, A. Mathai, C. Murphy, S. Rahmatallah and A. Patrick. "Santos: A physics-based digital human simulation environment." In Proceedings of the Human Factors and Ergonomics Society Annual Meeting, Vol. 50, no. 20, pp. 2279–2283. Sage CA: Los Angeles, CA: SAGE Publications, 2006.
3. N. Badler. "LiveActor: A virtual training environment with reactive embodied agents." Workshop on Intelligent Human Augmentation and Virtual Environments, University of North Carolina at Chapel Hill, Oct. 2002.
4. A. Bloomfield, Y. Deng, P. Rondot, J. Wampler, D. Harth, M. McManus and N. Badler. "A taxonomy and comparison of haptic actions for disassembly tasks." IEEE Virtual Reality Conf., Los Angeles, CA, March 2003, pp. 225–231.
5. A. Bloomfield and N. Badler. "Collision awareness using vibrotactile arrays." Proceedings of IEEE Virtual Reality Conference, March 2007, pp. 163–170.
6. L. Zhao and N. Badler. "Acquiring and validating motion qualities from live limb gestures." Graphical Models 67(1), 2005, pp. 1-16.
7. C. Cruz-Neira, D. Sandin, T. DeFanti, R. Kenyon and J. Hart. "The CAVE: Audio visual experience automatic virtual environment." Comm. of the ACM 35(6), 1992, pp. 64-72.
8. SIGGRAPH '92 Showcase and CAVE® Documentation - Part 2. 1992. https://youtu.be/S8pBnu B5rys?si=DtPC4u1S1DPxGN-U
9. https://www.youtube.com/watch?v=TYHIk2OfmaU

Jack Joins the Military

27

Although we had conceived of *Jack* as an ergonomics and human factors model, its real-time interactivity opened doors to the world of military simulations. A key factor was my participation in Penn's 1984 large-scale U.S. Army Research Office Research grant for "Artificial Intelligence Research." Now we had regular exposure to the Army sponsors and highlighted our *Jack* work in numerous venues, workshops, and meetings. Word spread quickly through other branches, and we soon found sponsoring partners in the Air Force and Navy.

Military simulations for vehicles and aircraft had been in existence for a while, but they lacked realistically modeled people. The original simulated video of the 1991 coalition battle of "73 Easting" during the Persian Gulf War was a reconstruction of that event from GPS and sensor data. Sketchy generic 2D depictions represented soldiers and prisoners. As compute and display capabilities improved, scaling up content and refining simulation modeling granularity down to the individual soldier level became desirable. A special U.S. government office was set up to coordinate efforts across the various branches: the Defense Modeling and Simulation Office (DMSO). DMSO produced specifications and standards to ensure scalability, interoperability, and extensibility, as simulations incorporated new weapon systems and additional active human participants.

The key to distributed interactive simulation (DIS) is the use of protocol data units (PDUs). DIS allows multiple computers to run independent but coordinated aspects of a large-scale simulation, obviating the need to do everything on one massive machine. The PDUs communicated *state* to entities across the network. For example, the relevant PDUs for a soldier might be *walk, run, kneel, drop to prone, crawl, rise to stand, aim gun, fire gun*, or be *dead*. Parameters specified target, speed, and direction. This was a nice control

© The Author(s), under exclusive license to Springer Nature Switzerland AG 2025   125
N. Badler, *On Raising a Digital Human*, Synthesis Lectures on Computer Science,
https://doi.org/10.1007/978-3-031-63945-6_27

system: a simulation engine issued PDUs to virtual humans who executed and animated the desired state changes and actions.

Jack became the ideal 3D virtual soldier. We changed his clothing to Army fatigues. We had a walking locomotion model, so forward travel and turns were straightforward. With *Jack's* real-time inverse kinematics, he could hold a gun with two arms and aim it in any desired target direction. The only animations we had to hand-build and add were a crawling motion, and the posture changes from standing to kneeling, kneeling to prone, and vice versa. The PDUs essentially issued motion commands through the *Jack* API. We could transition from any state PDU to any other by animating arcs in a *posture graph* [1]. For example, to transition from a *walk* to a *crawl*, *Jack* would stop, go prone, and then crawl. The posture graph ensured that all transitions made sense and pushed the performance burden onto *Jack* rather than the PDU issuing agent.

The culmination of our military simulation effort was an integrated demonstration for the 1994 Interservice/Industry Training, Simulation and Education Conference (I/ITSEC). We collaborated with Professor Mike Zyda and colleagues at the Naval Postgraduate School in Monterey, California, and Sarcos Robotics in Salt Lake City, Utah [2] (Fig. 27.1). Sarcos built a VR platform called iPort so that the user could virtually navigate the 3D environment and see it through an immersive VR display. The field of view in the VR helmet was controlled so that the forward direction was always centered even if the iPort was turned. A servo constantly and gently rotated the iPort back to the neutral forward position. The user was unaware of this subtle adjustment. Of course, the point of the demo was to find and track down the PDU-controlled *Jack* virtual humans who appeared, ran, hid, and shot at the trainee.

We could control several *Jack* models through PDUs. We did our best to make illustrative videos of platoon-size activities, but we were not military experts and could only demonstrate potentialities. Nonetheless, we had a deliverable product and saw it used by our sponsors. In 1993, the Naval Air Warfare Center Training Systems Division (NAWCTSD) began embedding *Jack* in its products. The sponsor expressed concern with a certain lack of physical realism in the soldier model and its movements. To remedy this, NAWCTSD contracted with a private company, Boston Dynamics, to custom-build and animate an alternate human model. That model was called DI-Guy. DI-Guy soon took over all the *Jack* military simulation implementations by providing a superior range of clothed models and better movements. The DI-Guy motions were actually motion-captured from soldier subjects. I found out some years later from Marc Raibert, founder of Boston Dynamics, that one of my former undergraduate students, Brian Stokes, ran the original DI-Guy motion capture sessions around 1994 at Lamb & Company in Minneapolis, Minnesota. Stokes had learned to use our Ascension motion capture system at Penn at a time when few such systems were available in university labs. More ironic than tragic, I eventually appreciated that Boston Dynamics did us a great favor by realizing the necessary virtual soldier technology transfer. Doing so kept us out of a costly maintenance and support mission inappropriate for a university lab. As a codicil to the DI-Guy evolution, I

Fig. 27.1 The iPort soldier interface. Three virtual *Jack* models are visible, each dressed in camouflage. On the left is the user's avatar, showing his position and pose in the interface. This avatar was movable anywhere in the scene by user swiveling and pedaling in the iPort. In the upper right two *Jack* "combatants" are visible targets for the military simulation. Image courtesy of Michael Zyda

only recently found out a further connection. Around 2002, MAK Technologies programmer Aline Normoyle integrated DI-Guy into a major military simulation framework. She left her MAK position in 2006 and enrolled in graduate studies at Penn, following her partner's appointment to the Wharton faculty. MAK eventually bought DI-Guy in 2013. In 2015, Normoyle became my last Ph.D. student.

I only saw *Jack* in an actual DIS setting once more. I no longer remember when or where only that it was a conference room. I recall watching a small 3D battle. A *Jack* soldier was standing a short distance away from a burning tank. I was talking to my host and not paying much attention to the screen since no one was issuing control commands. Suddenly, the *Jack* soldier took several steps away from the tank. "Oh," my host said, "he started to feel the heat from the fire and moved away." I knew *Jack* was a computer graphics model, I knew the fire was an animation, and I knew *Jack* did not have any reasoning ability. But at that moment all that I knew to be true was vanquished by the sheer naturalness and human-like response *Jack* showed to an external stimulus. The era of building spontaneous human behaviors into simulated beings had begun.

References

1. J. Granieri, J. Crabtree and N. Badler. "Production and playback of human figure motion for visual simulation." ACM Transactions on Modeling and Computer Simulation 5(3), July 1995, pp. 222–241.
2. D. Pratt, P. Barham, J. Locke, M. Zyda, B. Eastman, T. Moore, K. Biggers, R. Douglass, S. Jacobsen, M. Hollick, J. Granieri, H. Ko, and N. Badler. "Insertion of an articulated human into a networked virtual environment." Proc. of the Conf. on AI, Simulation and Planning in High Autonomy Systems, University of Florida, Gainesville, December 1994, pp. 84–90.

In 1974, the medical imaging community changed forever with the publication of the first Visible Human dataset. Sponsored by the National Library of Medicine, the University of Colorado had sectioned the entire frozen body of an executed prisoner into 4 mm thick 256×256 spatial resolution slices [1]. Although voxel displays had been around for some time to recreate CT imagery in 3D, there was no anatomical ground truth to compare any specific scan against an atlas or standard model. The visible human male was not a perfect specimen; for example, he had injuries from the lethal injection and lacked a testicle. The preparation method would be refined for thinner slices and higher resolution in the second dataset of a Visible Woman.

Voxels storing human anatomy sound great, but voxel models have resolution and connectivity issues. For example, attempts to trace the circulatory system fail when the capillary size falls below voxel resolution. Structures do not neatly align along voxel dimensional boundaries. Human variability makes automated anatomical matching between a real patient and the Visible Human models tricky even if everything is normal, and complex when other structures are present or absent and anthropometric variability is factored in.

As is well understood in computer graphics, having a shape model of something does not imply that we know its function. The Visible Human could possibly be used to infer function, but the voxel model itself was not conducive to readily transforming anatomy into physiology.

In 1994, Colonel Richard Satava of the Defense Advanced Research Projects Agency (DARPA) invited me to a workshop on the Visible Human. He had apparently heard about our work on real-time *Jack* for military simulations. Thinking I would just be an observer, I did not bring along my usual thick notebooks of 35 mm slides from which

© The Author(s), under exclusive license to Springer Nature Switzerland AG 2025   129
N. Badler, *On Raising a Digital Human*, Synthesis Lectures on Computer Science,
https://doi.org/10.1007/978-3-031-63945-6_28

I could customize a talk. Nevertheless, I did what I could when I was surprised by his call for me to speak, and I described our work with real-time people for simulations. Ultimately, the question Col. Satava posed to me was whether *Jack* could have something like Virtual Human innards and, even better, physiological models that could represent injury and interventional processes.

This was in perfect alignment with the work I had begun with Professor Webber. She already had an ongoing collaboration with a top Philadelphia trauma surgeon, Dr. John Clarke. Some of their interests lay in assessing potential internal organ injury paths given the set of observed entry and exit wounds on a shooting or stabbing victim. Professor Dimitris Metaxas added expertise in physical simulation, and with mutual students, we had our first publications on linking anatomy and physiology within a year [2].

Between 1994 and 1996, with the National Library of Medicine (NLM) and Col. Satava's DARPA support, we integrated physiological models for respiration and blood loss into the real-time *Jack* figure using Becket's PaT-Nets. These were an enormous help for modeling the simultaneous physiological state changes, appearance manifestations, and action animations on the victim's body. Dr. Clarke gave us invaluable professional medical advice in setting up the physiological models. Ph.D. students Diane Chi, Evangelos Kokkevis, Lola Ogunyemi, and Rama Bindiganavale, along with *Jack* software specialist Mike Hollick, created a real-time injury simulation and management system. The intention would be to show how computer graphics and VR could aid "golden hour" victim stabilization for emergency battlefield medic training. We named this integration "MediSim" [3]. MediSim became the Center for Human Modeling and Simulation's "go-to" demo.

The interactive MediSim system presented a user with a simulated *Jack* casualty lying supine (face up) on the ground. He was initially clothed with no obvious surface wounds (for example, he had all his limbs). A second *Jack* figure portrayed the animated medic (Fig. 28.1). He would do nothing until actions were selected from the available menu. The simulation operator could select the victim's injury, and then the physiological models would begin to function. Now the trainee would take control of their avatar medic. A possible course of action would be to do nothing, in which case the victim usually died in about three minutes if the selected injury was fatal (We sped up the physiology to make the process more exciting.). A good injury (so to speak, for simulation purposes) was a tension pneumothorax or punctured lung. If untreated, the victim turned blue, began labored chest movements, lost blood pressure, and had distended neck veins. The intervention is to insert a chest tube to relieve the pressure on the lung and restore a decent level of respiration. From an on-screen menu, the user gave the simulated medic commands to make tests, such as heart and respiration rates, and perform actions, which were executed in real time and affected the inputs to the physiological models. These actions triggered stored animations. We punted on difficult moves such as removing the victim's shirt; it just disappeared. The medic had a kit that he used (somewhat schematically) to access various tools or tests. MediSim may have been the first instance of two virtual people interacting in the same

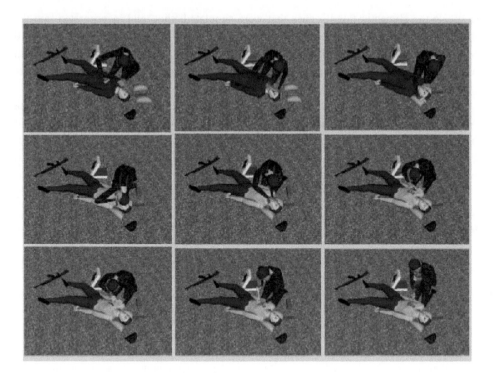

Fig. 28.1 Selected frames from a MediSim session

space with nontrivial simulated consequences. The MediSim demo was so compelling that a user could not tolerate just walking away without trying to stabilize the victim! Better to lose him than to ignore his plight.

Early in the MediSim project, I started investigating microscale physiological properties. In particular, I thought it might be necessary to concern ourselves with blood clot timing and bleeding for the simulation. One of the project Ph.D. students, Nick Foster, began to study blood flow behaviors. During the summer of 1995, he created a low blood flow and clotting model with 2D fluid simulation. The SIGGRAPH Conference that year was in Los Angeles, and several of us from Penn attended, including Foster. The Conference Reception was held in Pershing Square. Foster and I walked into the reception together, passing by a programmed water jet fountain near the entrance. I joked with Foster, saying something to the effect of: "Wouldn't that be great to do with computer animation? Not much difference between leaking blood and shooting fountain jets." After returning to Philadelphia, Foster continued with his thesis studies under Dimitris Metaxas, with an emphasis on 3D fluid flow. Metaxas had much better credentials than I in the physics simulation space. Foster graduated in 1997 with a Ph.D. thesis on 3D gas and fluid flow simulation and joined Pacific Data Images while they were in production on the movie *Antz*. Foster began work on their 3D fluid simulator. He completed it in time to create the climactic flood scenes that were simulated, shot, and incorporated into the film. In 1998, Foster received a Motion Picture Academy Technical Achievement Award for his software system. There was a short but traceable trail to his success starting with MediSim. Unfortunately, after contributing animation tools to subsequent films, Foster passed away in 2020.

Toward the end of our grant from the NLM, we transitioned MediSim to Sandia National Labs, where development and further integration with military simulations would be led by Dr. Sharon Stansfield [4]. Stansfield had been a Ph.D. student at Penn under Professor Bajcsy some years earlier. I judge that transition to be a success story. The best research is sometimes that which you can let go.

References

1. National Library of Medicine. "The Visible Human Project." https://www.nlm.nih.gov/research/visible/visible_human.html . Accessed November 11, 2023.
2. D. DeCarlo, J. Kaye, D. Metaxas, J. Clarke, B. Webber, and N. Badler. "Integrating anatomy and physiology for behavior modeling." In *Interactive Technology and the New Paradigm for Healthcare*, K. Morgan, R. M. Satava, H. B. Sieburg, R. Mattheus and J. P. Christensen (Eds.), IOS Press and Ohmsha, 1995, pp. 81–87.
3. D. Chi, E. Kokkevis, O. Ogunyemi, R. Bindiganavale, M. Hollick, J. Clarke, B. Webber and N. Badler. "Simulated casualties and medics for emergency training." In *Medicine Meets Virtual Reality*, K.S. Morgan, H.M. Hoffman, D. Stredney, and S.J. Weghorst (Eds.), IOS Press, Amsterdam, 1997, pp. 486-494.
4. S. Stansfield, D. Shawver and A. Sobel. "MediSim: A prototype VR system for training medical first responders." Proceedings, IEEE Virtual Reality Annual International Symposium, 1998, pp. 198–205.

Eyes Alive

As our human models improved over the years, we were not alone in encountering the "uncanny valley" aptly named by roboticist Masahiro Mori [1]. As artificial representations of humans become more accurate, they reach a point—the uncanny valley—where they become creepy. People are exquisitely tuned to making these "not quite human" visual assessments, probably for good evolutionary "friend or foe" reasons. While *Jack*'s segmented turtle-like torso and lack of good skin deformation algorithms were clear visual flaws, they kept the models out of the uncanny valley. When our graphics improved, their most disconcerting feature became their "deer in the headlights" wide-eyed stare. Addressing this problem led us on a fruitful computational adventure with four Ph.D. theses.

Catherine Pelachaud's work on faces first introduced us to the complexities of eye simulation. The muscles around the eyes form and control a range of emotional responses. Pelachaud realized that she needed to incorporate eye blinks. She did this procedurally, by executing blinks with appropriate regularity but also critically synchronized with stressed syllables during speech production. Facial Action Coding instructions controlled eye gaze directions, so they were programmable but not autonomously generated.

We returned to the eye gaze issue with Ph.D. student Sooha Park Lee. She used an eye tracker to determine the distributions of gaze directions and timings of eye saccades during conversational states such as speaking, listening, and idling. She used these distributions to parametrically control an animated character's eye movements during a simulated conversation. These saccades—fast, short eye displacements—added a "liveliness" to the character animation that was surprisingly realistic. Fortuitously, during 2001, my son Jeremy was working on his Ph.D. in eye movement neuroscience at the University of

N. Badler, *On Raising a Digital Human*, Synthesis Lectures on Computer Science, https://doi.org/10.1007/978-3-031-63945-6_29

Leuven in Belgium. Jeremy became our scientist co-author on the SIGGRAPH publication [2]. Many years later, Jeremy facetiously lamented the fact that, as a neuroscientist, his most frequently cited paper was in computer graphics! His scientific contributions were recognized again when he helped author a 2014 Eurographics survey on gaze and eye animations [3].

Saccades form an appropriate "background noise" behavior for eyes during a dyadic (two-person) conversation. Eyes could return to a forward gaze to appear engaged with the partner. However, in reality, eyes pay attention to the entire visual environment. We had been trying to model directional eye gaze driven by environmental motion. One motivation came from maintenance training, where the simulated technician had to look at the object components, observe possible hazards, and aim at various reach targets. Ph.D. student Sonu Chopra-Khullar created an attention model that directed eyes to moving objects in the scene [4]. The model was active, in that it built a trajectory predictor for any movement. Eyes did not need to track something continuously. The predictive model led to human-like behaviors if a visible object became occluded. The model predicted the time and location the object should emerge from behind the occluder, and would "look" there at the predicted time, expecting its reappearance. The model would be "surprised" if the expectations were unmet. Babies usually display this gaze behavior before one year of age [5]. The model could also sample objects for anticipated changes of state: for example, the human would periodically look at a traffic light for the color change that would indicate that it was safe to cross the street—after looking both ways for oncoming traffic!

I thought our final effort on eye gaze problems would be Erdan Gu's 2006 Ph.D., which combined conversational eye movements with environmental distractors in a multi-agent (group) setting [6]. Once again, Jeremy was our scientific anchor. Constructed with a turn-taking algorithm to distribute utterances across a group of virtual participants, their eye gaze was mediated by their speech, their engagement with the speaker, their desire to speak, and even by environmental distractions. Computer vision algorithms detect distractions in real time directly from a video stream [7]. My favorite demonstration of this arrangement showed a student sitting at a workstation with a webcam, talking with Pelachaud's Greta digital human model [8] (Fig. 29.1). While he is talking, another student walks behind him. The computer vision algorithm notices the moving distraction, and Greta shifts her eye gaze to follow the student until she leaves the screen space. This effect is both surprising and disconcerting since we expect the character to be paying attention to the speaker and not the background environment. Nonetheless, it is certainly a human response to the situation. In today's technological world of real-time avatars, webcams, Zoom meetings, and digital people, our virtual interlocutor may appear extremely invasive if it is distracted by and reacts to other people, pets, or events in our personal video background. Because it is visually "right" may not make it ethically "right." Perhaps it is better to know that the virtual being saw what it did than have it observe, record, but not

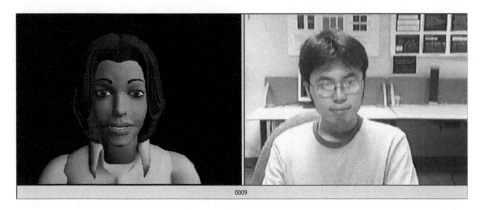

Fig. 29.1. While the student on the right talks to the virtual Greta female figure (on the left but actually on the screen in front of him), her gaze is captured by another student who passes the webcam behind the student

overtly react. Like using the "Men in Black" memory erasure gizmo, at least we'll know what we want to delete.

Whenever I thought we had finished with faces, I was proven wrong. We returned to eye gaze behaviors when Professor Soraia Musse at the Pontifícia Universidade Católica do Rio Grande do Sul in Brazil sent us one of her Ph.D. students, Vini Cassol. He would do his required Ph.D. "sandwich" year at Penn. Cassol took up the eye gaze issue, working with ideas and impetus from another Penn Ph.D. student, Aline Normoyle. Again, Jeremy was our convenient neuroscientist. Together with an undergraduate, Teresa Fan, we investigated whether eye gaze alone could influence how a player interacts with a Non-Player Character (NPC) [9]. They used NPC eye movements to convey "trust"—how the NPC felt toward the player. Trust was probabilistically modulated as the proportion of time the NPC looked at either the player or at a peripheral target. One would expect that greater trust would be paired with more eye contact with the player. We found that viewers did associate NPC gaze differences with trust and not with an unrelated attitude such as aggression. The effect held for different facial expressions and scene contexts as well. As humans have evolved to read other people for social cues, it is hardly surprising that the same interpretative mechanisms appear to work when encountering computer-generated characters.

References

1. M. Mori. "The Uncanny Valley: The Original Essay." June 2012. https://spectrum.ieee.org/the-uncanny-valley/ .
2. S. Lee, J. Badler and N. Badler. "Eyes alive." ACM Transactions on Graphics - Special Issue, Proceedings of SIGGRAPH 2002, San Antonio, TX, pp. 637–644.

3. K. Ruhland, S. Andrist, J. Badler, C. Peters, N. Badler, M. Gleicher, B. Mutlu and R. McDonnel. "Look me in the eyes: A survey of eye and gaze animation for virtual agents and artificial systems." Eurographics STAR (State-of-the-Art Report), 2014.
4. S. Chopra-Khullar and N. Badler. "Where to look? Automating attending behaviors of virtual human characters." Autonomous Agents and Multi-agent Systems 4(1/2), 2001, pp. 9–23.
5. R. Woods, T. Wilcox, J. Armstrong and G. Alexander. "Infants' representations of 3-dimensional occluded objects." Infant Behavior and Development, 33(4), Dec. 2010, pp. 663–671.
6. E. Gu, S. P. Lee, J. Badler and N. Badler. "Eye movements, saccades, and multiparty conversations." In *Data-Driven Facial Animation*, Z. Deng and U. Neumann (Eds). Springer-Verlag, London, 2008, pp. 79–97.
7. E. Gu, J. Wang and N. Badler. "Generating sequence of eye fixations using decision-theoretic attention model." Workshop paper, IEEE Computer Vision and Pattern Recognition, 2005.
8. I. Poggi, C. Pelachaud, F. de Rosis, V. Carofiglio and B. De Carolis. "Greta. A believable embodied conversational agent." In *Multimodal Intelligent Information Presentation*, O. Stock and M. Zancanaro (Eds.), Springer, 2005, pp. 3–25.
9. A. Normoyle, J. Badler, T. Fan, N. Badler, V. Cassol and S. Musse. "Evaluating perceived trust from procedurally animated gaze." Proc. Motion in Games, 2013.

Parametrized Action Representation

30

Jack's departure to Transom in 1996 was a relief, but it also created a vacuum. I could refocus on new research directions and not just oversee code and customer feature developments. The ANIML animation from instructions work was moving ahead with our Air Force support. In a general redirection of effort, we could now work at so-called "higher levels" of control where *Jack* "people" were the target of other driving mechanisms. We could exploit PaT-Nets for gesture *Jack* conversations, for Medisim simulations and physiology, and for maintenance task execution and error detection. PaT-Nets themselves were API targets for AI planning systems and cognitive models. We could study agents with increased autonomy who were capable of navigating a 3D environment and interacting with each other without following a strict script. A talented team of several Ph.D. students from the ANIML seminar built PaT-Nets and a multi-room environment so that several *Jack* people could autonomously play a legitimate game of hide-and-seek, including finding hiding places and trying to reach home before being tagged [1]. We were making some progress in connecting natural language and human animation.

In 1998, I had the proverbial rug pulled from under me when I learned that my colleagues Bonnie Webber and Mark Steedman were leaving Penn for Chaired faculty positions at the University of Edinburgh in Scotland. Webber and I were collaborators for almost two decades. Through ANIML and MediSim, we had mentored and fostered a community of students with a strong awareness of intellectual problems spanning language, simulation, and graphics. Steedman and I collaborated on simulating faces, speech, and emotion, primarily through shared students and postdoc Catherine Pelachaud.

I was too deep into these projects to brood over any feelings of abandonment. I could only wish Webber and Steedman well in their new roles. They were still personal friends, and they kindly invited me to visit and give a talk in 1999 at Edinburgh where they were already pursuing their own research directions. It was time for me to seek a new natural language collaborator if the language components of my work were to enjoy intellectual credibility.

As Webber and Steedman were leaving Penn, they promoted the idea of hiring Dr. Martha Palmer as a new faculty member in CIS. Palmer was already working on the Penn campus as a Research Fellow and Director of Technology in the Institute for Cognitive Science. She had a Ph.D. from Edinburgh and a credible publication record. Her research focused on verbs since at least 1990 [2]. Since human actions are often describable by verbs, verbs could be the basis of continued collaboration for the ANIML students. Palmer became a Visiting Associate Professor in CIS in 1998 and an Associate Professor in 1999. She could now supervise Ph.D. students.

Palmer and I restarted a regular seminar attended by many of our Ph.D. students working across language and animation. To further mutual interests, we agreed to focus on verb representations that could drive human animations. From a language perspective, verbs had discrete parametric slots for tense and argument structures. From the animation side, verbs were a potential descriptive avenue for creating motions if we could transform those slots into motion parameters. We already had PaT-Nets that could relieve language-level control of some low-level, body-specific details. We needed an in-between layer to connect these two ends.

We decided to name this in-between layer *Parametrized Action Representation* or PAR. It would be a representation, in the classic AI sense of the term, with named fields and knowable values. PAR was about actions and thus related to motion verbs. PAR would be parametrized to facilitate connections to lower-level graphics animation methods. We now had to define what PARs looked like and develop computational transformations to connect them with language and graphics components. We imagined four schematic components. The "downward" animation direction was first mapped from language and verb expressions into PAR, then the second mapped from PAR to graphics controls. Equally interesting was the reverse "upward" direction, where graphics animation could be recognized and stored in the PAR format, then output by generating natural language sentences from PAR. This was my Ph.D. thesis again but based on much-updated techniques.

We believe PAR was the first comprehensive attempt to connect language and animation through a tangible representation. One of Webber's remaining Ph.D. students at Penn, Julie Bourne, made the first pass on the PAR definition leading to her 1999 Ph.D. thesis. Other students modified and expanded that framework for the first announcement in 1998 [3] and subsequent book chapter publication by 2000 [4]. These students would be a cohesive group to cover most of the downward and upward transformations enabled by PAR.

Motion verbs by themselves are not descriptively sufficient for animation. Our LambdaMOO space used verbs that triggered straightforward state changes like "sit," or predetermined procedural motions such as "go to." Movement richness, subtlety, and even emotional content are enhanced by adverbs and prepositions. We already had a

Fig. 30.1 Initial PAR
Specification

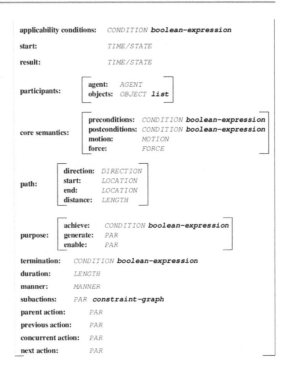

applicability conditions: *CONDITION* **boolean-expression**

start: *TIME/STATE*

result: *TIME/STATE*

participants: | agent: *AGENT*
               | objects: *OBJECT* **list**

core semantics: | preconditions: *CONDITION* **boolean-expression**
                 | postconditions: *CONDITION* **boolean-expression**
                 | motion: *MOTION*
                 | force: *FORCE*

path: | direction: *DIRECTION*
       | start: *LOCATION*
       | end: *LOCATION*
       | distance: *LENGTH*

purpose: | achieve: *CONDITION* **boolean-expression**
          | generate: *PAR*
          | enable: *PAR*

termination: *CONDITION* **boolean-expression**

duration: *LENGTH*

manner: *MANNER*

subactions: *PAR* **constraint-graph**

parent action: *PAR*

previous action: *PAR*

concurrent action: *PAR*

next action: *PAR*

handle on certain adverbial modifiers through the EMOTE system and its application to American Sign Language synthesis [5]. Ph.D. student Jugal Kalita looked at how prepositions augmented verbs with spatial context [6]. Linguist Annette Herskovits had already produced a major study of English spatial prepositions [7], but we were specifically interested in how we might animate prepositions of movement. Working with Ph.D. student Yilin (Dianna) Xu, we built computational realizations for all of them [8]. We discovered that the most interesting movement prepositions began with the letter "a": *along*, *around*, and *across*. The shape of the target object heavily influenced the paths these prepositions encapsulated. The role of context in computing spatial relations, especially "near," had been investigated previously by others [9], but we saw contextual influences in movement paths as well. Xu went on to complete her 2002 Ph.D. under Professor Jean Gallier, but in my biased opinion, her preposition work was her most significant contribution.

Our first exposition of PAR laid out its essential features. These included and better codified some aspects of PaT-Nets (Fig. 30.1).

PARs were organized in a hierarchic directed graph. A PAR could have a *parent* or a graph of *subactions*. PARs could be linked to a *previous* and *next* action and have explicit *concurrent* actions. *Applicability conditions* tested whether or not a PAR was relevant to a situation, avoiding the need to search through and evaluate the full set of PARs in particular circumstances. A *termination* condition likewise caused it to finish. A PAR referenced the *objects* involved and the instigating or active *agent*. Agents were either virtual people or objects with self-motive capability, such as a vehicle. *Path* coordinates or directions could be stored, and actions could be generated either kinematically as a motion or dynamically by exerting a force. A *purpose* included explicit conditions for achieving a goal, generating another action, or simply enabling the possible execution of some action. *Manner* allowed the specification of LMA/EMOTE movement modifiers for individual variation in a generic motion. The choice of representational slot names sympathized with language parsing and generation requirements, yet also linked, as PaT-Nets had, to animatable motions in the *Jack* Toolkit.

The term *agent* reflected our emergent view that we were no longer in the body modeling business. We wished to create active, participatory, and observant human agents, literally "smart agents." Their action vocabulary would be stored in a specialized database of PARs. I coined the term *Actionary* for this structure. I really liked the term and considered trademarking it for some time, but didn't. Instead, Joshi, Palmer and I submitted an NSF project proposal titled "The Actionary" and received 3 years of funding starting in 1999.

As we had many times before, significant effort went into finding an appropriate demonstration vehicle. We wanted to showcase our overall vision as well as emphasize the particular individual contributions of students across the NL/PAR/Actionary/*Jack* spectrum. We realized that we could leverage work we started on a previous grant in 1998 from the Office of Naval Research (ONR) on "Developing Virtual Environments for Training" or VET. Our partners were *Jack*'s owner at that time, Engineering Animation, and a colleague from my NASA collaborator days, University of Houston Professor Bowen Loftin. He had achieved some fame building VR training environments for NASA missions and science education [10]. The general scenario we finally agreed on was a virtual environment trainer for a vehicle checkpoint (Fig. 30.2). Checkpoints were an essential component of access control during both conflict and peacetime. They could be relatively simple configurations of barriers and steel drums suitable for cover, but they required crucial human elements: there were at least three people, a driver and two or three checkpoint control officer trainees. Access depended on the first officer properly assessing the situation and gauging the emotional state of the driver. The other trainees were backups in case of emergency. We named this demonstration system VET after the ONR project.

Three Ph.D. students led the VET effort: Rama Bindiganavale, Jan Allbeck, and William Schuler. Palmer and Joshi cosupervised Schuler, who handled the natural language to PAR compilation. Allbeck built the Actionary and PARs. Bindiganavale produced the motion generators [11]. PAR's manner field allowed specifying the behaviors of the driver to show their emotional state. The key feature offered by the explicit structure of the Actionary and the PARs was to allow online *modification* to behavior rules in natural language. For example, in an early training scenario, the first trainee draws his weapon at a suspected terrorist driver but forgets to take cover. The driver shoots him. Wrong trainee behavior. To correct this, the user gives the following *standing order*: "When you

Fig. 30.2 Virtual Environment for Training (VET): *Jack* models in the checkpoint scenario

draw your weapon at the driver, take cover from the driver behind your drum." Standing orders parsed from the sentence become rules (that is, PARs) that are duly applied if their preconditions are met, in this case, triggered by the presence of a driver with a weapon. A survey paper on language-to-animation systems in 2016 attested to PAR's generative power [12].

In addition to generating PARs from natural language, we looked at human motion trajectories for PAR construction. Bindiganavale's 2000 Ph.D. combined motion capture with some of the latent concepts from my Ph.D. thesis on motion chunking to develop a methodology for learning PARs from performed examples [13]. She hypothesized that acceleration zero-crossings were key motion events. A zero-crossing occurs when the movement of some body part, notably the hands, changes from accelerating to decelerating, or vice versa. Starting to move a hand toward a target causes a positive acceleration, and as it approaches the goal, it must begin to decelerate. This positive-to-negative transition is a cue that a goal has been achieved, and a PAR is created to describe the action. By using motions captured from a live subject or synthesized by a *Jack* model, Bindiganavale's program could learn most of the fields of a PAR from *one example*. Unlike typical machine learning methods that require multiple training cases, Bindiganavale's method behaved more like humans do: show it just once what you want, and it can probably do a good job of reproducing the task even if the reach target, human model, or grasped object identity changes. Reaching for, picking up, and drinking from a cup is a prime example of how this process achieves movement generality across agents. Even though arm length, stature, or vessel can change, the agent learns it has to keep it level and bring it up to touch its lips.

We continued to evolve details in PAR beyond the initial structure. The PAR field for manner terms provided an explicit hook to connect verb definitions, motion capture, and our EMOTE work. In his 2001 Ph.d. thesis, Liwei Zhao trained neural networks to recognize the occurrence of significant EMOTE values from a live human motion capture performance [14]. These fed the PAR manner field directly. Since we could interpret and reanimate the values in the manner field through EMOTE, we had control over the visible performance of an otherwise generic PAR motion. A few years later, in 2008, Ph.D. student Durell Bouchard would close another gap in PAR generation by determining when live-captured human motions might be segmented into successive PARs by the detection of significant EMOTE parameter changes rather than just acceleration zero-crossings [15]. Bouchard had generalized my Ph.D. thesis technique of basing action chunks on trajectory changes to triggering *human* action units from EMOTE.

References

1. T. Trias, S. Chopra, B. Reich, M. Moore, N. Badler, B. Webber and C. Geib. "Decision networks for integrating the behaviors of virtual agents and avatars." IEEE Virtual Reality Annual International Symposium, 1996.
2. M. Palmer. "Customizing verb definitions for specific semantic domains." Machine Translation 5, 1990, pp. 45–62.
3. N. Badler, R. Bindiganavale, J. Bourne, M. Palmer, J. Shi and W. Schuler. "A parameterized action representation for virtual human agents." In Workshop on Embodied Conversational Characters, Lake Tahoe, CA. 1998.
4. N. Badler, R. Bindiganavale, J. Allbeck, W. Schuler, L. Zhao and M. Palmer. "Parameterized Action Representation for virtual human agents." In *Embodied Conversational Agents*, J. Cassell, J. Sullivan, S. Prevost and E. Churchill (Eds.), MIT Press, 2000, pp. 256–284.
5. L. Zhao, K. Kipper, W. Schuler, C. Vogler, N. Badler and M. Palmer. "A machine translation system from English to American Sign Language." Proc. Association for Machine Translation in the Americas, 2000.
6. J. Kalita and N. Badler. "Interpreting prepositions physically." AAAI-91, Anaheim, CA, July 1991.
7. A. Herskovits. "On the spatial uses of prepositions." In 18th Annual Meeting of the Association for Computational Linguistics, Philadelphia, PA. Association for Computational Linguistics, 1980, pp. 1–5.
8. Y. Xu and N. Badler. "Algorithms for generating motion trajectories described by prepositions." Proc. Computer Animation 2000 Conference, IEEE Computer Society, Philadelphia, May 3–5, 2000, pp. 33–39.
9. Danovsky, M. "How near is near?" MIT AI Laboratory Memo AIM-344, Feb. 1976.
10. C. Dede, M. Salzman and B. Loftin. "ScienceSpace: Virtual realities for learning complex and abstract scientific concepts." Proceedings of the Virtual Reality Annual International Symposium, VRAIS '96, Los Alamitos, CA, IEEE Computer Society Press, 1996, pp. 246–252, 271.
11. R. Bindiganavale, W. Schuler, J. Allbeck, N. Badler, A. Joshi and M. Palmer. "Dynamically altering agent behaviors using Natural Language instructions." Autonomous Agents 2000, pp. 293–300.

12. K. Hassani and W.-S. Lee. "Visualizing Natural Language descriptions: A survey," ACM Computing Surveys, Vol. 49 (1), Article No. 17, 2016, pp. 1–34.
13. R. Bindiganavale and N. Badler. "Motion abstraction and mapping with spatial constraints." Workshop on Motion Capture Technology, Geneva, Switzerland, November 1998.
14. L. Zhao and N. Badler. "Acquiring and validating motion qualities from live limb gestures." Graphical Models 67(1), 2005, pp. 1–16.
15. D. Bouchard and N. Badler. "Semantic segmentation of motion capture using Laban Movement Analysis." Proc. Intelligent Virtual Agents (IVA), 2007.

When *Jack* moved out of Penn in 1996, Transom took along two of my staff programmers, Mike Hollick and John Granieri. Granieri would continue in the Ph.D. program and obtain his degree in 2000. Meanwhile, I had to find a new lab manager since the position opened with Granieri's departure. I tapped one of the veteran ANIML students, Jan Allbeck, to take over. Allbeck would not have the deep *Jack* code responsibility accorded to Granieri and Hollick, so we could direct her energies in new directions.

Allbeck took on the new lab manager position with some trepidation. She had a quiet assertiveness but always tempered by humility, honesty, and humor. She had been only peripherally involved in the *Jack* code, but I assured her that she could rely on her predecessors and Transom to handle most of those issues. I was hoping Allbeck could move us ahead with agent modeling. One of her first assignments involved reading Neal Stephenson's classic novel *Snow Crash* and writing a critique of its futuristic vision compared to 1998 technical reality [1]. Allbeck did not disappoint. Over the next ten years, in addition to managing the HMS Center, she would be our lead technologist in agent behavior modeling.

Allbeck and I felt that PAR had the potential to control complex multi-agent human activities, but we were not yet utilizing its full power. Allbeck appeared interested in exploring the "multi" aspect of agent modeling: that is, the patterns and activities that involved many human agents acting and interacting together. In 2002, she led a broad-based post-ANIML effort to animate multiple agents with her ACUMEN system [2]. ACUMEN attempted to both create and describe small crowd behaviors with language terms derived from LMA movement parameters. This problem soon split into two directions: one on large-scale pedestrian crowd behaviors and Allbeck's own interest in a *functional* population.

© The Author(s), under exclusive license to Springer Nature Switzerland AG 2025
N. Badler, *On Raising a Digital Human*, Synthesis Lectures on Computer Science,
https://doi.org/10.1007/978-3-031-63945-6_31

What distinguishes a crowd from a functional population is that the former typically have navigation and destination goals, while the latter have *additional* meaning, purpose, responsibilities, and schedules. Crowds are often associated with homogeneous mass movements and generic goals, such as pedestrians navigating a city street, transit riders utilizing a train station, or occupants exiting a building during an emergency. Functional populations exhibit heterogeneous behaviors, such as what transpires within an office building, school, or entertainment venue, specifically during normal, expected, or non-emergency activities.

An excellent opportunity to pursue functional crowds arose in 2007. Led by my CIS colleague Professor Mitch Marcus, he and I received a 5-year "Multidisciplinary University Research Initiative" (MURI) grant from the U.S. Army to combine natural language, animation, and robotics. Allbeck would be my lead on the animation aspects. She would also liaison with the robotics group through one of their Ph.D. students, Hadas Kress-Gazit.

Allbeck categorized agent behaviors into four types and accordingly chose to name her system CAROSA: Crowds with Aleatoric, Reactive, Opportunistic, and Scheduled Actions. *Aleatoric* actions were random within well-defined sets with probability distributions; for example, what gestures a teacher might display during a lecture. They won't be random but would conform statistically to some set of likely gestures that accompany speech. Different action distributions would accompany other classroom activities, such as quiet reading or grading. *Reactive* actions were triggered by detected conditions, a basic component of the PARs that underlay the animated behaviors. *Opportunistic* behaviors "wait in the background" until conditions make it possible for them to execute. For example, having to use the restroom may have to wait until the end of class, or the opportunistic passing of a restroom on the way to some other desired target. *Scheduled* actions occur when a calendar event triggers them.

The user interface is one of the smartest features of Allbeck's CAROSA implementation. She decided that popular, readily available, and semantically adequate interfaces could be used to establish each agent's behavior. Her user interface used Microsoft Outlook® to hold agent responsibilities, personal calendars, friend lists, and task obligations. These were linked to PARs so that when a behavior arose, say from a calendar entry, it could be sent to the agent for animation. She also incorporated a resource manager. For example, students entering a classroom were allocated seats from the available set. Preassigned seats were unnecessary. If no seats remained, they could stand in the back or turn around and leave. Allbeck completed her Ph.D. in 2009 after co-authoring our first book on crowd simulation [3].

As Allbeck was wrapping up her CAROSA work, the flexibility of the MURI grant allowed us to study an alternative view of how functional crowds might be created. Instead of the *agent-centric* perspective of CAROSA, where every agent had tasks, needs, and a calendar, Catherine Stocker collaborated with Allbeck and others to propose an *event-centric* view. Stocker's *smart events* were scheduled or requested, but which agent responded to the request could vary according to a number of *extrinsic* parameters, such as who was close to the location of the desired event, who had the capability to achieve the goals of the event, and who was available [4]. This approach worked well for "accidental" situations. For example, if a fire started in a room, the event required an agent to douse the fire. While all nearby agents would receive the same notification, the agent actually responding to the call would be someone who was trained to fight the fire, who knew where the fire extinguisher was kept, and who was available to approach and extinguish it. These events were on nobody's *intrinsic* calendar, but each agent was *primed* with abilities and capabilities that could be used as a filter to select one or more actors. The beauty of this arrangement is that it appeared to the observer that agents had actual agency—that is, that they were making decisions that affected their environment. In fact, it was the other way around: the events in the environment co-opted available agents to perform the required actions. Similar to the animation example I cited earlier, where a door's own movement drives the reach point of the opener's hand, smart events drove agents to perform more complex, multi-person, coordinated tasks.

The distinction between a virtual human model and an animated agent is not only a matter of shape and appearance. When we imbued the *Jack* human model with inverse kinematics, its ability to reach and follow a target interactively afforded it an aura of cooperativity. It could *assist* the human factors engineer with problem solving. With the CAROSA system, collections of animated human models could appear to follow humanlike behavior regimens based on schedules and needs. They assumed identifiable roles, behaving like teachers, students, office workers, or restaurant customers. These appearances belied the fact that they were still automata, controlled by PARs that dictated what to do and when. Smart events shifted control to events and removed most actual decision-making to agent selection processes. Our agents appeared to be in control of their actions, even if that appearance was a fiction.

Appearances do matter, and to understand why I need to return to Michotte's classic psychological studies. Michotte's animations involved completely abstract geometric figures such as large circles and little squares. He completely side-stepped any uncanny valley issues since these geometries resembled no animate forms. Yet subjects invariably attributed causality, and even personality, to the animated shapes. The perception of agency lies in the observer. The veracity of cartoon animation builds on this visual experience. Cartoons also avoided the uncanny valley, and animators learned how to portray actions that evoked animacy through movement alone [5].

One of our smart event demonstration videos drove home the notion of agency as an artifact of appearance. The demo scene takes place in a main room with some surrounding spaces connected through open doorways. Some agents stand in the main room, and other agents linger in the side rooms. A fire starts in a corner of the main room. One nearby agent is co-opted to sound the fire alarm. Other agents in the main room must exit. The alarm signals a firefighter with an extinguisher to enter the main room from one of the side rooms and approach the fire. During the animation of this scenario, one of the exiting agents heads toward the door where the firefighter is entering. Their paths accidentally align, and they both stop in the middle of the main room before escaping the deadlocked near-collision and continuing on to their respective goals. But what it *looks* like is that they stopped during their chance meeting to briefly *converse with one another*! While nothing in any representation of the scenario needed, allowed, or even considered this possibility, our own mental models provided the perception of causality consistent with plausible, situated, and explicable human behaviors. Apparently, we humans are experts at inventing explanations, causality, and attributions when in fact the situation is accidental, coincidental, or just random. Interpretation tempers the reliability of "eyewitness" accounts.

References

1. J. Allbeck and N. Badler. "Avatars á la Snow Crash." Proc. Computer Animation, Philadelphia, PA, June 1998.
2. J. Allbeck, K. Kipper, C. Adams, W. Schuler, E. Zoubanova, N. Badler, M. Palmer and A. Joshi. "ACUMEN: Amplifying Control and Understanding of Multiple Entities." Autonomous Agents and Multi-Agent Systems, Bologna, Italy, July 2002.
3. N. Pelechano, J. Allbeck, and N. Badler. *Virtual Crowds, Methods, Simulation and Control.* Synthesis Lectures on Computer Graphics and Animation, Morgan & Claypool, 2008.
4. C. Stocker, L. Sun, P. Huang, Q. Wenhu, J. Allbeck and N. Badler. "Smart events and primed agents." Intelligent Virtual Agents (IVA) 2010.
5. J. Lassiter. "Principles of traditional animation applied to 3D computer animation." ACM SIG-GRAPH Computer Graphics, 21(4), July 1987 pp. 35–44.

SIGGRAPH and Service

When I began my faculty position at Penn in 1974, I did not completely give up on my Ph.D. interests, and for a while, I kept up with computer vision publications. I left most of this field to my colleague Professor Ruzena Bajcsy. As a parting effort to maintain a role in computer understanding of motion, I organized and Chaired an IEEE Workshop on Computer Analysis of Time-Varying Imagery, held in Philadelphia in the spring of 1979. I subsequently served as a Guest Co-Editor (with Professor J. K. "Jake" Aggarwal of the University of Texas at Austin) for a 1980 Special Issue of *IEEE Transactions on Pattern Analysis and Machine Intelligence* (PAMI) on "Computer Analysis of Time-Varying Imagery." While I was happy to help launch a new direction in image processing, computer vision did not excite me as much as I thought it would. I was anxious to return to my 3D graphics roots and embedded myself in a new community: SIGGRAPH, the ACM Special Interest Group in Computer Graphics.

Shortly after joining Penn, in 1975 I started co-teaching a computer graphics course with a Wharton Lecturer, Tom Johnson. In 1976, Johnson volunteered to Chair the ACM SIGGRAPH Conference on the Penn campus. When the initial Tutorials Chair could not receive the essential work release to plan the conference events, Johnson asked me if I would step in and take over. Thus, I became the 1976 SIGGRAPH Conference Tutorial Chair. This was the break of a lifetime. That year, we had four tutorials with speakers such as Ed Catmull and Tom DeFanti. These people were bringing exciting new concepts to computer graphics, such as shaded rendering, 3D modeling, animated art, and analog video image processing. SIGGRAPH' 76 had the first hardware show, held in the roof lounge of one of the Penn high-rise dormitories. Maxine Brown, a computer science Master's student, organized the nascent hardware exhibit.

© The Author(s), under exclusive license to Springer Nature Switzerland AG 2025   149
N. Badler, *On Raising a Digital Human*, Synthesis Lectures on Computer Science,
https://doi.org/10.1007/978-3-031-63945-6_32

Both Brown and I continued to volunteer for the next few years, following the respective organizational paths we blazed. I was part of the core SIGGRAPH community leadership. In 1979, I ran for SIGGRAPH Vice-Chair and won. I was re-elected in 1981 but resigned a year early in 1982 due to mounting research obligations at Penn, such as our TEMPUS work for NASA. During those 3 years, SIGGRAPH Conference attendance ballooned from about 3,000 to 14,000, almost an order of magnitude increase [1]. The SIGGRAPH Executive Board formed the Conference Oversight Committee, which I led, and finally offloaded Conference Management duties to a commercial enterprise, Smith-Bucklin. For several years, I co-taught the Introduction to Computer Graphics Course (Courses were the new name for the Tutorials) until handing it over to Andrew Glassner.

My "last gasp" attempt to foster closer collaboration between my interests in both computer graphics and computer vision was to broker an arrangement that allowed discounted cross-registration between the 1981 SIGGRAPH and the Pattern Recognition and Image Processing (PRIP) conferences being held simultaneously in Dallas, Texas. I was on good terms with the Chairs of both: Tony Lucido of SIGGRAPH and Jake Aggarwal and Azriel Rosenfeld of PRIP. I believed that vision methods had utility for graphics, and that the detailed models developed for graphics could inform better recognition methods for vision. Although history would later validate my intuitions, the experiment was an abject failure, with maybe one cross-registrant, possibly just me.

Azriel Rosenfeld deserves more than a passing mention. As both Ruzena Bajcsy and I were interested in computer vision, around 1980 we invited him to give a talk at Penn. Rosenfeld was the de facto leader of the computer vision community. He founded the journal *Computer Graphics and Image Processing* in 1972. I had already published a couple of papers in his journal. During his visit to Penn, he invited me to join its Editorial Board. That was a nice break for me, a young, just-tenured Associate Professor. By 1982, he promoted me to senior co-editor, alongside well-known vision researchers Linda Shapiro, Thomas Huang, Herbert Freeman, and himself. In 1983, the journal changed its name to *Computer Vision, Graphics and Image Processing*. In 1990, the journal recognized the massive growth in each of these fields by splitting into two. The graphics half became *Graphical Models and Image Processing*. University of Maryland Professor Rama Chellappa and I became its senior co-editors. In 1990, we renamed it again to just *Graphical Models*. Ingrid Carlbom and were its co-editors-in-chief through 2009. Although we thought that *Graphical Models* would be an apt title for a computer graphics journal, we realized too late that "graphical models" was a phrase used in probability theory and in the nascent machine learning community. I take the blame for this clash. I wish I had Internet search available to check this in 1990.

Although my research interests came first in my career, the SIGGRAPH human network and the wider publication community were crucial in forming the myriad professional relationships that I enjoy to this day. The most tangible results have accrued to my students—from undergraduates to Ph.D.s—as these networks sometimes helped them find satisfying technical careers in animation, special effects, social networks, games, and programming. The citation for my induction into the SIGGRAPH Academy in 2021 explicitly recognized my mentorship. To signify that honor, a ceramic replica of the iconic computer graphics teapot graces my bookshelf.

Reference

1. https://en.wikipedia.org/wiki/SIGGRAPH

Teaching

I never had an overt personal goal to teach. I was not a particularly extroverted young man, but even in high school, I had no social issues with giving stand-up presentations or running for student office. I was elected the Student Council Chair, and my best friend, Steve Abbors, was the Senior Class President. My role models were the teachers who nurtured my sometimes unusual interests, whether in building a binary computer in grade school or allowing me to design a kinetic sculpture instead of writing an essay for my senior year English class project. I was fortunate to have good math and science teachers in the 1960s public schools. Other teachers did damage, perhaps without even knowing it. My junior year English teacher told me I couldn't write. Rather than trying to fix that problem, he caused a mental block against writing—hence a kinetic sculpture in senior-year English. Years of university classes could only partially reverse the stigma. I didn't learn to write decent prose until I co-authored technical papers at Penn with colleagues Steve Smoliar and Bonnie Webber. They had professional text editing experience and covered my drafts with red ink. I learned how to write "on the job."

I think my decision to attend UCSB and then enter the College of Creative Studies was key to observing quality university teachers at work. I am not saying that professors at higher-ranked schools can't or don't teach; they may have been hired for their research prowess. UCSB had a research reputation, but teaching abilities appeared to be a prominent faculty attribute. The CCS mathematician, Professor Max Weiss, taught through on-the-board experience. Other CCS mandates allowed me to take interesting courses and not just requirements. I escaped freshman English by taking two-quarters of Russian language. In sophomore year, I took a modern algebra class at 8 am; Professor Adil Yakub was so enthusiastic and dynamic that he always kept me awake. CCS had a visiting computer professional, David Culler, teaching computer concepts that ignited my

© The Author(s), under exclusive license to Springer Nature Switzerland AG 2025   153
N. Badler, *On Raising a Digital Human*, Synthesis Lectures on Computer Science,
https://doi.org/10.1007/978-3-031-63945-6_33

interest in graphics. I took a course on modern art and loved it. I then took a course on modern architecture from another visiting faculty. She almost converted me to study architecture in graduate school! Fortunately (for me), I had already committed to attending the University of Toronto. I'm certain I would have made a very poor architect.

The University of Toronto professors were good teachers as well as strong researchers, and I had to pay attention as I was compressing my computer science coursework into just 2 years. The computer industry barely existed in 1974. The only obvious direction for a Ph.D. was research and teaching. Although I taught some courses in my final year at Toronto to earn a living, when hired at Penn I had to learn to be a serious teacher. This took nontrivial effort. One was (at that time) simply expected to walk into the classroom and know what to do, as if one gained such skill through osmosis. My main task, early in my career, would be teaching computer programming. For the most part, this was straightforward since the introductory course was about skill development and programming homework. I could do that, I had Teaching Assistants, and together we did the job.

Soon, though, I proposed teaching a specialized introductory programming course in Penn's "General Honors" category. The idea was accepted, and I taught this for a few years. Unlike the existing introductory course, which was populated mostly by engineering students and computer science majors, the General Honors version catered to non-technical Arts and Science students prequalified for that program. I could emphasize both learning to program and completing an *individualized* programming project. The latter was the exciting part because I could tailor the project to a student's other interests. For example, a linguistics student might write a KWIC indexer for a document. There were some unexpected revelations: one art student rebelled against the rigid design syntax and structure of programming and wanted to write much more freeform "artwork" code—not a good approach. I liked this course because of the individualization. Perhaps I was channeling Max Weiss.

I started teaching computer graphics in 1975, at first with Wharton Lecturer and SIGGRAPH 1976 Conference Chair to be, Tom Johnson. After Johnson left Wharton, I taught computer graphics by myself. In 1978, when the only other Penn Engineering faculty member interested in CAD failed to get tenure in Civil Engineering, I could immediately fill the resulting vacuum. By the 1980s, I was also teaching part of the course "Introduction to 3D Modeling and Animation" at the SIGGRAPH Conference, so I had incentives to develop—or copy—good teaching materials. But by then my research fully occupied my time, and I knew my teaching took a back seat.

One of my SIGGRAPH Executive Committee colleagues, Vice-Chair Steve Levine, invited me to help him teach professional, week-long computer graphics courses. He had too many to do them all by himself. I signed on and, as a condition of employment, I had to watch about four hours of videotapes on "how to teach." For me, it was the first time someone codified any part of the teaching process itself. I subsequently taught many of these short courses, though they were a grind. The money was good, and the audiences were mostly attentive, though some were there because they were required to attend. One of the last ones I gave was at a government office outside Washington, DC. It seemed like computer graphics was an odd fit for their mission, so I asked at the end what they were thinking of doing with their graphics knowledge. "Oh, nothing," they replied, "We're required to learn about graphics since we wanted to purchase a color printer."

Computer graphics is a fun visual topic to teach. During the early 1980s, I developed extensive notes handwritten in colored marking pens on letter-sized clear plastic sheets called viewgraphs. Every classroom had a viewgraph projector. I could walk into class and show the prepared viewgraphs rather than laboriously draw diagrams on the chalkboard. By this time, too, I heartily disliked chalk dust, so the viewgraphs were healthier, too. I distinctly recall a watershed event rather early in one term. I was just getting into the graphics diagrams using my prepared viewgraphs when I looked down at a student in the front row. He was frantically trying to copy my drawings into his notes. He had purchased a special 20-color barrel pen, so he could rapidly switch colors and faithfully reproduce my drawings. At that moment, I promised to hand out the notes in a printed packet ahead of time. Producing, printing, and distributing notes in advance of lectures was the best stylistic teaching change I made.

Not long thereafter, as I refined my viewgraphs, I started to "prettify" them by converting all the hand-drawn materials to PowerPoint and then photographing them onto 35 mm slides. This took most of a summer, but I then had preloaded slide carousels I could carry into class and project with my own slide projector. Early technology in classrooms was a working electrical outlet, if lucky. This is not a joke. To show animation I needed to check out a 16 mm movie projector. Once, while showing a graphics movie, the lights went out. All the outlets were on a single circuit, so when someone turned on the coffeemaker in the Chemical Engineering department office down the hall, it tripped a circuit breaker.

I could print out my PowerPoint slides as nice notes all bound in one document for the start of class. Teaching computer graphics became a process of updating, correcting, and expanding these notes each year. I eventually ended up with hundreds of slides in several chapters. I learned about many emergent graphics topics as I developed the teaching materials. With the advent of installed video projectors in classrooms, I could finally dispense with 35 mm slides. I illustrated many graphics algorithms by visually reproducing their steps with PowerPoint animations. Now, too, I could just put the slides online and avoid the print copy. This was such an extensive slide collection that when one of my Ph.D. students graduated and desired to go into teaching, I would offer them my slide sets as a parting gift.

I was teaching well enough, but I still gave it a second academic priority. I came to realize two things. The first is that the purpose of teaching is to inspire learning. I wasn't just pouring facts into a funnel; the concepts had to be motivating, exciting, and useful. Second, I was at my best when I was giving talks about our research programs, especially on virtual people. I was talking to people about something exciting to me and hopefully to *them*, and I was able, apparently, to make that excitement infectious. Oddly, most of this research material never found its way into formalized Penn coursework, although the SIGGRAPH courses did provide a teaching outlet. People were willing to put money into my game, as it were, and see something tangible become of it. I became decently good at motivating why we were doing our research, and recounting enough history to provide context and background. I started giving keynote talks: the sort that are meant to frame a conference theme, contrast existing approaches, or inspire cross-connections and new ideas. Between 1982 and 2021, I was invited to give 26 Keynote or Plenary talks at both domestic and international venues across Asia and Europe. The audiences also varied. In addition to computer graphics and animation-themed conferences, they included virtual rehabilitation, digital human modeling, and emotional understanding.

Through almost 50 years of teaching, I never quite escaped the impostor syndrome. I had nagging doubts about my teaching effectiveness. My teaching evaluations were adequate. I felt that I did best in courses with individualized projects. These often broadened my own experience. It was rewarding to see students achieve on their own. I felt more success in non-credit seminars such as ANIML, where the topics had evolving novelty, the rewards were theses and papers, and there were no grades. My Ph.D. students were like family. To watch their intellectual growth, to find their own topics, to feel pride in their accomplishments, and to see them pursue their own careers has been (almost) as satisfying as having one's own children.

Digital Media Design

My post-*Jack* years were a time of reflection and redirection. Research on agent modeling, action representation, and cognitive capabilities was ongoing, but my pool of Ph.D. students began to shrink with graduations. On the teaching side, Bonnie Webber and I had experimented with teaching an "Interactive Systems" course in the early 1990s which combined aspects of natural language and computer graphics interfaces with interactive devices. Soon thereafter, one of our graphics lab Ph.D. students, Jeffry Nimeroff, took it over and turned it into a Virtual Reality course.

One day, Nimeroff came to my office and observed that our various computer graphics and interaction courses seemed to be sufficient to support a new "Master's Program in Virtual Environments." We were inspired by Professors Randy Pausch and Don Marinelli, who founded the Entertainment Technology Center Master's program at Carnegie Mellon University in 1998. I definitely liked this focus, and I encouraged Nimeroff to formalize his new Master's program proposal. The plan that Nimeroff produced implied that he could be the Program Director if the eventual enrollment was sufficient. I would be the faculty advocate and try to shepherd it along the approval path.

The CIS faculty agreed with the Master's in Virtual Environments plan, and the engineering school approved it as well. There was one more hurdle: as a new graduate program, Penn's Board of Trustees had to review it. In a private university, the Trustees are its literal owners. I presented the proposal to the Trustees, and they voted their approval. Great! I would be its initial director. I was ready to get to work. I am slightly fuzzy about the immediate next details, but the following year, we were extremely disappointed to have very few, or maybe even no, applicants for the program.

© The Author(s), under exclusive license to Springer Nature Switzerland AG 2025   157
N. Badler, *On Raising a Digital Human*, Synthesis Lectures on Computer Science,
https://doi.org/10.1007/978-3-031-63945-6_34

Although initially unbeknownst to me, our engineering school dean, Greg Farrington, saw an opportunity in that failure. Farrington proposed recasting the program as an *undergraduate* degree. Moreover, he would seek the collaboration and support of two other deans in the Annenberg School of Communication and the Graduate School of Fine Arts. Legend has it that the three deans walked into the Provost's office and refused to leave until he approved it. Farrington even renamed it in appreciation of his colleagues' support. He called it "Digital Media Design" (DMD), using a term associated with each School: Engineering, Communication, and Fine Arts. Somewhat unfortunately, no one bothered to check that Penn already had a "DMD" degree: a *Doctor of Medicine in Dentistry*. While context should readily disambiguate between the engineering DMD and the dentistry DMD, the clash has caused some confusion among potential students, and especially their parents.

DMD would be an engineering degree, though intellectually supported by three schools. The core steering committee for DMD comprised Julie Saecker-Schneider in the Graduate School of Fine Arts, Paul Messaris of the Annenberg School, and myself, as Director, in CIS. DMD went into the Penn undergraduate application in 1998 as a checkbox along with the other Engineering School majors and degrees. University rules forbid any outside advertising of undergraduate programs, so we just had to hope interested and eligible students would read the application description. In 1998, DMD took internal Penn transfer students, and in 1999, it admitted its first full class of students. We received funding from the Annenberg School for an assistant, David Phillips, who was a postdoc there at the time.

In 1999, David Phillips took a faculty position at the University of Toronto. The manager for Academic Affairs in the School of Engineering, Joseph Sun, told me about a person he knew in the Penn Admissions Office, Amy Calhoun, who might work out as a staff replacement. I interviewed her and hired her immediately. Amy became the de facto face of DMD until her retirement from Penn in 2023. Enormous credit for DMD's success is due to her engagement in the student application reviews, course quality assessments, on-site advising, and career placement of our DMD students.

Our normal DMD senior class size ranged from 10 to 20 students, so we were considered a "boutique" degree in Engineering. In essence, DMD defined what a computer graphics undergraduate degree should look like. DMD was a computer science degree, and our students took the same technical courses as the CIS majors. The main degree differences were our DMD requirements for computer graphics courses in engineering, and for fine arts courses in drawing and 3D modeling. For many years, we also required a communications course from Annenberg. We felt that those were foundational skills for someone going into the graphics field. The DMD application required some sort of creative portfolio. Our students would soon swell the class sizes in those fine arts courses, and we received great support from their faculty, especially Instructor Scott White and Professor Joshua Mosley.

To launch the DMD degree, I had to convert Nimeroff's VR course and the interactive systems course I previously offered with Webber into something novel. Given the exploding popularity of computer games, I called the hybrid offering "Interactive Game Design." I co-taught this sophomore-level course until the mid-2000s with various colleagues, including Joshua Mosley from fine arts, Paul Diefenbach (now on the faculty at neighboring Drexel University), and a computer graphics friend from the Princeton area, Dr. Stephen Lane.

I think this course was initially successful because it became a socialization venue where our DMD students could get to know one another. This was, of course, before online social media took over. It also gave us a course where DMD students could help as teaching assistants and exploit their own game-related experiences. Using undergraduates as teaching assistants was a new and rather scary concept for the CIS faculty, who expected to use Ph.D. students in this capacity. The notion of nearly-peer mentorship in class further cemented DMD social bonds.

We could give this class small team assignments and an individualized final game project. I adapted a course idea I first saw in Professor Don Greenberg's Cornell computer graphics course where the students had to create 3D models from some classic piece of 2D art. We expanded this into creating a *game* where the artwork itself appeared as the visual theme at the game onset or materialized at a successful conclusion. The student teams each chose 1 of 50 possible starting art images. It was great fun to see what creative inventions were derived from these pictures.

Only after teaching a course for a few years do all its positives and negatives come into focus. The game design courses produced fun interactive artifacts, but I realized that we were emphasizing creative expression over technological competence. Nothing wrong with the former, but we were failing on the latter. Our DMD engineers were not acquiring the programming skills necessary to compete as *technologists*. Accordingly, I made a hard turn and replaced the entire game design course with a brand-new version that contained C++ programming instruction, Qt interaction interfaces, and 3D graphics fundamentals. We called it simply "Interactive Computer Graphics." I relied on a co-instructor for the C++ and Qt components, while I repurposed much of my extant computer graphics course materials into a series of programming projects leading to a substantial final project.

To jumpstart the introduction of Autodesk Maya into DMD, I invited its co-developer and University of Toronto Professor Karan Singh to teach a 3D modeling course at Penn for a semester. After exploring a few programming project options for my introductory graphics course, one of the DMD TAs, Matthew Kuric, hit on what we called "Mini-Maya": a 3D object modeling system with a number of basic but crucial Maya features. All the DMD students had to learn Maya in their fine arts 3D modeling class. Now they had to understand *how* it worked and program their own version! As my colleague Steve Lane aptly remarked, they had to learn what was going on "under the hood." Soon I stopped teaching this course and handed it off to a competent former DMD student and now Senior Lecturer, Adam Mally. For a few years, some of the graphics content was co-taught by Lecturer Dr. Benedict Brown until he left Penn for Yale.

This course change produced tremendous intellectual success. We educated computer graphics students who could do serious programming. There were two primary unanticipated consequences. First, computer science students discovered that the interactive computer graphics course taught them C++ and certain software engineering skills missing from other required CIS courses. The second turned out to be the most important from a personal perspective. Because our DMD students could program *and* had creative abilities, they began to collaborate on our research projects. From the mid-2000s onward, undergraduate students began to appear as co-authors on our research papers. DMD became a research asset: they could make complex 3D models, attractive animations, and help implement supportive algorithms. Paul Kanyuk and Sunny Teich collaborated on their DMD senior project to implement "distribution maps," which scattered trash-like 3D artifacts in a scene depending on the probability expressed in the map. When Teich graduated in 2005, she joined Weta in New Zealand to work on their forthcoming movie *Avatar*. Kanyuk distinguished himself as the first non-Ph.D. student hired into the research group at Pixar. He soon became their crowd Technical Director for movies such as *Ratatouille* and *Wall-E*. Undergraduate co-authors appear on at least 24 of my publications. They also collaborated with other faculty, such as Professor Chenfanfu Jiang, while he was at Penn. They understood how to make compelling stories and excellent visuals.

DMD students took internships and full-time positions in the animation industry as a new kind of student: not just an artist and not just a programmer. They could interpret artistic criteria and translate them into usable interactive tools. Even in non-industry settings, our students made an impression. Penn has an active Study Abroad program where students can spend a semester away at other prestigious international institutions. Several DMD students opted for the University of Otago, New Zealand. I had a respected colleague there, Professor Geoff Wyvill, who loved the DMD brand. He gave them interesting projects, and they enjoyed the professional and cultural change-of-pace. I was disappointed when Wyvill retired and my students could no longer experience his mentorship and humor. They could still go elsewhere, however, to find interesting and challenging opportunities.

DMD put me on the Engineering School's administrative radar. I had already served as CIS Department Chair in the 1990s, so I understood how the School "worked." Engineering Dean Eduardo Glandt asked me if I would become the Associate Dean for Academic Affairs. I agreed. I would have oversight of all academic levels in the Engineering School: undergraduate, Master's, Executive Master's, and Ph.D. degrees. The position came with a professional manager, Joseph Sun, and significant and competent staff.

I ceded my already spacious office in the Moore Building former's ENIAC space to Amy Calhoun and moved into one of the largest faculty offices on the Penn campus. Although it was palatial, its size didn't necessarily reflect power. It was supposed to be two regular-sized offices, but the renovation to split it into two would have been awkward.

It had an inside door and an external door leading to a hallway. The staff could monitor access through the interior door, but my own students and colleagues could circumvent the reception area by knocking on the hallway door. By this time, I had accumulated a large collection of books, reports, papers, and teaching materials. I had walls of bookshelves and six filing cabinets. I also had a portable video cart with a VHS tape player and a large TV monitor for teaching graphics, since none of the classrooms had video equipment in 2001. I had only been in this office for a few months when 9/11 happened. Since I had a TV, the staff and I watched the unfolding tragedy in my office that day.

DMD students were a joy to work with because they had numerous creative interests besides engineering and mathematics skills. Since DMD was launched just before I became the Associate Dean of the Engineering School, I was in a unique position to oversee its progress and impact alongside the other majors. DMD admitted and transferred about 20 students a year: 4–5% of the engineering freshman class. One of my favorite tasks as Associate Dean was giving the "Come to Penn!" pitch to visiting families of admitted students during the so-called Penn Previews campus visit week. Lunch with faculty and representative students culminated the event. With some current DMD students, Calhoun and I would work the lunch tables occupied by admitted DMD students and their families. We could answer questions firsthand and make our final pitches to clinch an acceptance. Because DMD was already a unique engineering program, most of the time the DMD students who came to Penn Previews had already accepted admission as early decision. It was great fun interacting with them and their families, anyway.

I clearly remember a conversation at one of these lunches. I was talking with a family whose son was accepted. His younger sister was along for the visit. It was her, and not the student-to-be, who asked the best question I ever heard at these events: "What makes DMD students special?" Although momentarily floored by the sheer brilliance of this question, I came up with an immediate answer: "DMD students are special because they are all different." By requiring some evidence of tangible creativity in the Penn application portfolio, we had ensured that our students would have the artistic and intellectual skills to tackle both the fine arts and computer science components of the degree. Their artistic explorations, however, assured individuality.

An unexpected but welcome side effect of the DMD degree was its attractiveness to women. We had not designed DMD with this specific goal in mind, but DMD's support of creativity made it seem like a more humane, less regimented, and even fun, engineering degree. With the dot-com boom of 1999 and the subsequent bust in 2000, women applicants dropped precipitously in many computer science undergraduate programs, including Penn's. By the early 2000s, however, women were the majority in DMD. I rationalized this statistic by naming it "The McDonald's Effect": if you ordered a hamburger meal, it was more satisfying with a side dish. Computer science was our "hamburger," and fine arts was our "side dish." The combination provided an attractive mental escape from the isolated programmer "cubicle" stereotype. The presence of DMD women apparently encouraged additional enrollments in computer science. Soon women accounted for at least 30% of the computer science class as a whole. I like to think that much of DMD's contribution to CIS at Penn lies in motivating, building, and cementing this diversity.

Joseph Sun came into my office one day and suggested that we try to start a pre-college summer engineering program. Not only would such a program expose high school students to engineering but it could also serve as a "feeder" for potential Penn under-graduate applicants. I thought this was an excellent idea. On Sun's recommendation, we decided that the pilot version would have one course: a compressed version of some of the DMD material on 3D modeling and animation. That next summer, Penn Engineer-ing offered a summer residence course in computer graphics with one of my Master's students, Mark van Langeveld, as the instructor. We paid stipends to some DMD stu-dents as Residence and Teaching Assistants. It was a sell-out, filling the available lab seats. Under the umbrella title "Summer Academic Academy in Science and Technol-ogy" (SAAST), in subsequent years, we expanded into other topics, such as computer programming, robotics, and bioengineering. Langeveld graduated from Penn in 2005 and began his Ph.D. in computer graphics at the University of Utah. He has taught Penn's computer graphics SAAST course since its inception. Indeed, as hoped, many DMD stu-dents had their start in the SAAST course. SAAST alumni include Adam Mally, current CIS Senior Lecturer and DMD Director since my retirement in 2021.

What happened to that Master's of Virtual Environments program? My colleague Steve Lane rescued it and recast it as a Master's program in "Computer Graphics and Game Technology" (CGGT). Under Lane's capable leadership, CGGT launched in 2002 and was a fast success. Soon, both DMD undergraduates and CGGT graduates were taking the same introductory and advanced graphics courses. Since initiating DMD in 1998, CGGT in 2002, and Ph.D.s since 1975, we count over 1000 Penn alumni in computer graphics.

The SIG Center Renovation

35

Jan Allbeck's 14-year tenure as Ph.D. student and lab manager in the Center for Human Modeling and Simulation had a lasting impression on our physical environment as well as our research productivity. In 2007, HMS received tremendous news. Tripp Becket and Diane Chi, two of the Ph.D. students from my lab who worked at Susquehanna International Group (SIG) in nearby Bala-Cynwyd, Pennsylvania, convinced the company to fund a total renovation of our physical space. We still occupied the original ENIAC area, but decades of partial remodels insufficiently modernized the space. Moreover, Professor Dimitris Metaxas left Penn for a position at Rutgers University, and we hired a fresh Ph.D., Alla Safonova, from Carnegie Mellon University to step into the computer graphics slot. Safonova wanted to use her start-up funds for a large motion capture space. To do that, we would have to gut the lab and totally rebuild it from the original walls.

The renovated space would be newly designated "The SIG Center for Computer Graphics." Kennedy and Violich Architecture of Boston produced a design that restored the original fourteen-foot ceilings and stripped the entire interior space. Because of ENIAC, the Moore Building is certified historic. Any changes to the exterior, such as our new windows, had to match the existing design. The SIG Center would be reconstructed as a large motion capture and workstation studio, a conference room, one office, and a kitchen. Many lab users felt that a kitchen—a sink, fridge, and microwave—was essential.

Other donors provided named spaces. While a Ph.D. student in my lab, Tripp Becket met his future wife, another computer science student, Dawn. Tripp and Dawn Becket funded the conference room. A former student who had taken my graphics class sometime in the 1980s, Ramanan Raghavendran, named the student work area. Harlan Stone graced us with the new large motion capture studio. Our connection with Stone is itself a longer story which I'll get to later.

© The Author(s), under exclusive license to Springer Nature Switzerland AG 2025   163
N. Badler, *On Raising a Digital Human*, Synthesis Lectures on Computer Science,
https://doi.org/10.1007/978-3-031-63945-6_35

Allbeck, as lab manager, became the liaison between the various renovation parties: the architects, the Engineering School, the furnishing manager, and the occupants. She made all the necessary day-to-day decisions. We moved everyone temporarily to a neighboring Engineering building. We had no choice, but that added equipment, support, and even neighbor issues. What had been a sleepy hallway in our swing space now supported an active lab community. Allbeck gets all the credit for making this renovation work.

The new SIG Center re-opened in 2009. In the Moore Building, the SIG Center became a focal point in the main entrance area immediately across from a small ENIAC museum display. We put glass windows along the foyer exposure so that visitors could see the graphics activity without having to enter the lab. A couple of large TV monitors ran demos and student interview videos that touted the DMD degree program. A large, but low-resolution LED display wall flanked the long-ramped hallway along one side. With motion sensors at both ends, the wall was visually responsive to any people passing by. We could even program it, at least until a power supply died and a decision was made to "pull the plug" rather than repair it.

The SIG Center became a stop for the prospective student tours in the engineering complex. Escorted by student volunteers, they said what they wanted about the SIG Center, even though they were often wildly incorrect. I once overheard the tour guide say that the Center "put *Jack* in a school bus and crashed him every day." We also heard that the Center is where we "made Pixar movie animations" (we didn't!) because we visibly displayed movie posters signed by alumni who worked on the films. To rescue some element of truth, we subscribed to a Penn tour feature that allowed someone to call a local phone number, enter a code, and hear a recording of me describing the SIG Center's foci. We also placed a QR code on our room number sign that led directly to our Center's homepage. Whether the visiting families believed (or remembered) what they heard on the tour, it was still nice exposure for our special Penn asset.

Crowds

As we began the new millennium, Allbeck's ACUMEN crowd work seeded new research directions. Although her own efforts turned to heterogeneous groups simulated through CAROSA, the more classical roles for crowds were environments where rather homogeneous pedestrians shared one or more spatial goals. Typical environments included emergency egress, mass flows into and out of transit stations, and pedestrian commuter traffic. I was aware of excellent crowd simulation efforts by colleagues such as Craig Reynolds, Nadia Magnenat-Thalmann, Daniel Thalmann, Ming Lin, Dinesh Manocha, and Stephen Guy. I didn't have any particular urge to compete with them.

Then, one day, I had a visitor stop outside my office door. Flashing a wide smile and speaking as rapidly as possible, she declared, "Hi, I'm Nuria Pelechano and I have a three-year Fulbright grant to work with you and get my Ph.D.!" How could I resist? Pelechano had just received her Master's degree from University College London with Professor Mel Slater, so her pedigree was strong, her enthusiasm boundless, and her funding a blessing. Pelechano's approach to crowd simulation had its roots in a *social force* model pioneered by Dirk Helbling [1]. People and the environment influence behaviors through a combination of pseudo-physical forces, such as attraction toward a goal, repulsion to avoid collisions with solid objects, and mutual avoidance to prevent people from bumping into one another. Pelechano wanted to explore several additional influences: the role of leadership in evacuation processes, the effect of communicating route blockage information, and the real-time VR experience of feeling as if one were actually present in a moving dense crowd. She began these studies in earnest in 2003 and quickly assumed a leadership role in this domain.

Pelechano's first study examined the effect of communicating blocked routes in a purely geometric search to exit a maze [2]. She proved, via simulations, that interagent

N. Badler, *On Raising a Digital Human*, Synthesis Lectures on Computer Science, https://doi.org/10.1007/978-3-031-63945-6_36

communication sped up the egress progress. Agents could inform one another that certain passageways were blocked. They did not have to waste time exploring dead ends. She also considered the effect of having a mix of knowledgeable leaders and followers. No leaders implied that everyone just relied on random searches. If only about 10% of the agents actually knew the entire maze layout, they could plan routes to an exit. The rest of the agents could just "follow a leader" and escape without blind route searching. These were good, sensible, empirically validated demonstrations that crowd egress was not purely a geometric path-finding problem: the agents themselves played an important role in the evacuation speed and success rate.

Pelechano then turned her attention to the social force model. She modified the force expressions and added additional terms for new influential features, such as pushing, impatience, panic, vibration reduction, and alternative goal selection [3]. This crowd model, called HiDAC, ran in real time. We could build a VR experience where someone felt that they were present in a moving crowd. *Presence* means that a VR participant feels that they are "truly" *in* an environment. Like the impression of agency, I experienced with the smart event demonstration, the sense of presence in VR arose when individuals in the crowd exiting a room stepped aside to avoid colliding with *me* [4].

We explored the VR sense of presence further by creating a "cocktail party" environment with tables holding 3D models of various canapé trays. Wearing an immersive helmet, the VR subject had to walk around, visit all the tables, and collect a canapé from each. Numerous virtual "guests" were already present in the cocktail party space and were milling about on their own. In one videotaped episode, our subject approached a canapé table when he caught sight of a virtual agent moving immediately in front of him. The real-world video clearly shows the subject instinctively taking a quick step backward to avoid a collision and utter, simultaneously, an audible "oops." The cocktail environment was quite a compelling experience.

In 2006, 3 years and 3 months after arriving at Penn, Pelechano fulfilled her own promise and completed her Ph.D. Determination and skills were clearly evident in her achievements. A collaborator in the Architecture Department, Professor Ali Malkawi, hired her as a postdoc for another year. They continued building evacuation studies while Pelechano published her results. Ultimately, Pelechano, Allbeck, and I co-authored the *second* computer graphics book on crowd simulation [5]. Soraia Musse published her Ph.D. thesis with advisor Daniel Thalmann in the first book [6]. I would eventually co-author a series of three more books on the subject [7–9].

As Pelechano neared completion of her degree, I admitted a new Ph.D. student, Ben Sunshine-Hill. Sunshine-Hill was a gamer at heart. He wanted to get a Ph.D., but Penn was not his top choice. When his partner received and accepted a Ph.D. offer from another Penn department, Sunshine-Hill joined HMS. I started him on a hard crowd problem: creating plausible "background" character paths based on actual but only partly known human movement flows within a building. Pelechano's agents usually had only a small number of possible exit goals. Allbeck's CAROSA required that each agent had a schedule and explicit goals, but constructing hundreds of these would be tedious.

To make realistic trajectory choices for a large population inhabiting a building during everyday activities, one had to track them somehow. It would be awkward (and quite invasive) to hook up office workers to physical sensors. We did not want to use actual cameras to follow individuals. I saw a workshop announcement focused on massive datasets. They offered participants a huge dataset of motion sensor triggers in a large office building occupied by the Mitsubishi Electric Research Laboratory (MERL) in Cambridge, Massachusetts. These data sounded ideal. It was clearly anonymous and had no individual tracking information per se. Sunshine-Hill investigated whether sensor hits could lead to plausible trajectory inferences. Using agent probability models and a transition network called a Markov process to move people through likely successive sensor activations, Sunshine-Hill could establish good individual trajectories [10].

We put this work aside, however, mostly because the data itself required real-world instrumentation. In the spirit of new computer graphics paradigms, we wanted to develop procedural—that is, purely computational—generation techniques for crowd animation. Generating plausible agent navigation targets would greatly help crowd scenario scripting. Sunshine-Hill's game interest would be key.

Sunshine-Hill enjoyed playing a hugely popular game, Rockstar's *Grand Theft Auto*. The game includes many non-player entities, such as characters and vehicles. These components add richness and a plausible urban context to the game. One day, while watching an armored car drive past his avatar, he decided to follow it to determine its destination. Soon he discovered that it was just on a loop, driving around on a fixed path with no destination at all! That was easy for the game designer, but not at all an accurate portrayal of reality. Sunshine-Hill followed a few other NPCs around and observed the same result: these characters had circumscribed and repetitious paths. The NPCs were locally realistic but globally nonsensical.

Sunshine-Hill's observations led him to a deep insight and a clever solution. An NPC required a goal or intention only if it was *watched or followed*. That is, an NPC pedestrian, if ignored, could simply disappear around the corner of a building. But, if followed, the NPC had to invent a *reason* for its behavior. Sunshine-Hill called this process of associating a plausible goal with a watched NPC, *alibi generation* [11]. Taking some cues from CAROSA, NPC agents may have approximately scheduled destinations. However, they do not act on these goals unless followed, when they would need an alibi to justify their observed location and direction. For example, a watched agent may adopt an alibi of going to lunch if the timing is around noon. They will plan a path to the nearest restaurant where they can enter and conveniently "disappear." At other times in the day, they may enter an apartment house or a store: anything that seems appropriate for the time and place. A statistical distribution of convenient disappearance points prevents the simulation from degenerating into repetitive behaviors. Moreover, by using the player's perceptual region as an NPC filter, Sunshine-Hill did not have to create simulated agents outside that visible area unless they were scripted to enter. This greatly reduced scenario simulation effort. Sunshine-Hill presented his alibi generation to rave reviews at one of the annual Game Developers Conferences. He has since forged his career as Lead AI Developer at Havoc.com.

Allbeck, Pelechano, and Sunshine-Hill focused our crowd simulation effort on the concepts of intention and function. People need a reason for their behavior. If supportive visual cues are lacking, or if obviously bizarre, then they compromise the viewer's sense of presence and engagement in the scene. I am not going to criticize the crowd simulation community for lack of realism when there are still a multiplicity of interesting and difficult problems to solve. Human perceptual sensitivity makes it easy to draw attention to odd behaviors when two characters try to pass one another without a collision or to wonder why an agent doesn't turn sideways to squeeze through a narrow opening. Populating an outdoor venue seems to be important for enlivening it, but what or why these people are present is more elusive. Variations in dress, stature, appearance, gait, and accessories help, but do not in themselves provide an explanation for their existence. Interestingly, the accessories worn or carried by individuals are an often overlooked clue to pedestrian purpose [12, 13]. Applications in emergency egress have the advantage of a motivating factor: people need to exit a building for safety. However, what motivates people to walk from one place to another? They need an alibi or purpose, so they appear to belong in the scene. They may be headed to work, restaurants, transit stops, or home, and they may stop and look into windows, chat with acquaintances, or enter or exit vehicles. They may belong outdoors if they work as police, food vendors, or construction crews. Ultimately, the presence of people is part of the broader behavioral fabric of society, and without a story, the situation will appear to lack motivation and purpose. That became our next challenge.

References

1. D. Helbing, I. Farkas T. and Vicsek. "Simulating dynamical features of escape panic." Nature, 407, pp. 487–490, 2000.
2. N. Pelechano and N. Badler. "Modeling crowd communication and trained leaders during maze-like building evacuation." IEEE Computer Graphics and Applications 26, Nov. 2006, pp. 80–86.
3. N. Pelechano, J. Allbeck and N. Badler. "Controlling individual agents in high-density crowd simulation." ACM/Eurographics Symposium on Computer Animation, 2007, pp. 99–108.
4. N. Pelechano, C. Stocker, J. Allbeck and N. Badler. "Feeling crowded? Exploring presence in virtual crowds." Proc. of PRESENCE 2007. The 10th annual International Workshop on Presence, Barcelona, Spain, October 2007, pp. 373–376.
5. N. Pelechano, J. Allbeck, and N. Badler. *Virtual Crowds, Methods, Simulation and Control.* Synthesis Lectures on Computer Graphics and Animation, Morgan & Claypool, 2008.
6. D. Thalmann and S. R. Musse. *Crowd Simulation.* First Edition, Springer, 2007. (Second Edition, 2012)

7. M. Kapadia, N. Pelechano, J. Allbeck and N. Badler. *Virtual Crowds: Steps Toward Behavioral Realism.* Morgan & Claypool, September 2015.
8. N. Pelechano, J. Allbeck, M. Kapadia and N. Badler. *Simulating Heterogeneous Crowds with Interactive Behaviors.* Taylor & Francis, 2016.
9. V. Cassol, S. Musse, C. Jung and N. Badler. *Simulating Crowds in Egress Scenarios.* Springer, 2018.
10. B. Sunshine-Hill, J.M. Allbeck, N. Pelechano and N. Badler. "Generating plausible individual agent movement from spatio-temporal occupancy data." Workshop on Massive Datasets, Nagoya, Japan, 2007.
11. B. Sunshine-Hill and N. Badler. "Perceptually realistic behavior through alibi generation." Proc. AIIDE 2010.
12. J. Maïm, B. Yersin and D. Thalmann. "Unique character instances for crowds." IEEE Computer Graphics and Applications, 29(6) November 2009, pp. 82–90.
13. E. Wolf and N. Badler. "The distribution of carried items in urban environments." Presence: Teleoperators and Virtual Environments J., MIT Press, Vol. 24(3), Summer 2015, pp. 187–200.

By 2010, I agreed to transition most of my MURI project funding to Jan Allbeck so she could continue her CAROSA work in her new computer science faculty position at George Mason University. Allbeck's departure left me with two gaps: the HMS lab manager position and new Ph.D. student research projects. One of my advanced Ph.D. students, Joe Kider, agreed to take on the lab manager position. Kider had been a Ph.D. student for several years, but his progress stalled a while for health reasons. He was close to being booted from the Ph.D. program, but I went to bat for him and assured the faculty that I would get him moving again. My decision to defend him was the right call. Although his eventual Ph.D. thesis on simulating natural decay processes did not involve digital humans at all, he contributed greatly to a positive and supportive atmosphere in the HMS Center, especially for undergraduate research. Kider would later prove to be a crucial partner in two of my modeling projects, which I will turn to later.

Finding grant funds is one of the most challenging aspects of running a good-sized research enterprise. We never seemed to be short on ideas or on students. I had been quite fortunate to enjoy both large grants and small ones and benefitted from loyal sponsors. Filling in gaps between projects could be nerve-wracking since I never wanted to have to tell my lab manager or a graduate student that I could no longer pay them. Sometimes we were lucky with external fellowships, such as Erignac's French government, Pelechano's Fulbright, and Allbeck's Ashton fellowships. The Ashton was rather particular, as it was limited to Engineering Ph.D. students under 35 years old, resident in Pennsylvania or New Jersey, and having all four grandparents born in the U.S.

Lockheed-Martin became one of my local "savior" sponsors. Rich Rabbitz worked in a New Jersey branch across the Delaware River from Philadelphia. At Lockheed-Martin, Rabbitz was in a position to investigate human models and VR systems for Navy and

N. Badler, *On Raising a Digital Human*, Synthesis Lectures on Computer Science,
https://doi.org/10.1007/978-3-031-63945-6_37

Coast Guard shipboard applications. In 1991, Rabbitz completed his master's degree in CIS, opting to write a thesis on 3D model collision detection. Arranging an industry-to-university-funded project was painful. These required tedious contract negotiations, as both sides were jealously protective of their respective intellectual property rights. Penn would not engage in any classified work, either. Nonetheless, Rabbitz and I managed to forge almost a dozen yearly contracts between 1986 and 2010. Although individually modest in scope and dollars, they keep us alive and functioning, especially during leaner government funding times. I greatly appreciate Rabbitz's loyalty and support for all those years.

Toward the end of the Lockheed-Martin sponsorship run, we finally figured out how to get our Penn contributions into their proprietary software. We arranged for one of their staff programmers, Eric Halpern, to bring a copy of his software onto a SIG Center workstation that only he could access. Kider organized a team to work with Halpern, consisting of Master's student Damian Slonneger and DMD-ers Matthew Croop and Jeremy Cytryn. Halpern could rewrite Penn code into Lockheed-Martin code without having to disclose or share anything protected. That last joint project added some nice features on top of the *Jack* model, such as sitting and getting up from a chair, and elaborating the smart object work we had begun years before but not added to the commercial *Jack* product [1].

Meanwhile, I wanted to continue the crowd work that Pelechano, Allbeck, and Stocker had begun, but all had since moved to new positions. One day, I received a phone call from Professor Randy Shumaker of the University of Central Florida's Institute for Simulation and Training (IST). He had submitted a large-scale proposal to the U.S. Army, but they had requested that he include someone with computer graphics human factors expertise. He asked if I would like to join the revised proposal. Of course! IST's reputation was strong, and I would be happy to help out. In 2010, I signed on as a co-principal investigator and shared a small portion of a 5-year Army Robotics Collaborative Research Alliance grant. I would be their lead in human-robot interaction.

This project required coordinating digital humans with robots, or at least with 3D animated models of robots. The human-robot interactions centered on team operations in populated environments. Unexpectedly, but almost simultaneously with hearing of the project award, I received a postdoc inquiry email from a recent Ph.D. student, Mubbasir Kapadia. Supervised by Demetri Terzopoulos, Kapadia had just graduated from UCLA. I knew Terzopoulos quite well through his research on computer graphics, animated agents, and meaningful crowds. Kapadia sounded like the ideal postdoc to lead the Army project. I had no hesitation in hiring him, and he joined the HMS Center in 2011. Given Kapadia's background, knowledge, pedigree, and strong publication record, I figured I could relax a bit while he led the project. I couldn't have misjudged the situation more: during Kapadia's two years as my postdoc, he pushed out almost a paper a month, effectively co-supervised a number of my Ph.D. students and visitors, authored several proposals, and kept me busy with all his new ideas.

Kapadia's core interests centered on the human-robot interaction space, with an emphasis on groups or crowds of agents motivated by spatial goals, collision avoidance, and planning. He brought welcome skills in the planning domain. He soon assumed a de facto supervisory role with three of my Ph.D. students: Alex Shoulson, Cory Boatright, and Pengfei Huang. Like Kider, Kapadia also embraced the availability and skills of undergraduate and Master's students in the lab and motivated them to form collaborative teams and produce publishable work.

As in any new research undertaking, finding an appropriate testbed was crucial. We had some flexibility within the broad framework of Army interests in human-robot teams. We decided to construct a small, generic, but representative Middle East-like marketplace. We could populate it with locals in appropriate garb, and animate them doing their shopping. A soldier avatar would walk about the scene on patrol, accompanied by a mobile robot. This realization required storytelling components and a new approach to organizing character animations since the real-time soldier actions would influence non-avatar behaviors. This was similar to a game environment with local NPCs. Some undergraduate and Master's students constructed a suitable marketplace scene, found appropriately dressed human models online, and began to script the NPCs.

Alex Shoulson stepped in to create the first significant extension of the PAR paradigm since Allbeck's departure. He designed and implemented ADAPT: the Agent Development and Prototyping Testbed [2]. *Parametrized Behavior Trees* (PBT) were the primary structures in this system. Essentially, PBTs wrapped PARs in a set of explicit programming language concepts for action selection, repetition, conditional execution, random selection from distributions, and parameter passing. ADAPT would presage, at least structurally, the dataflow visual programming language style that came to dominate computer graphics user interfaces by the late 2010s in tools such as UnReal Engine's *Blueprints Visual Scripting*. Using ADAPT, we could easily give each agent a shopping list that required them to visit selected stalls. Once positioned at a stall, they could converse, perform gestures, and even haggle with the seller. They navigated the marketplace from stall to stall, avoiding collisions. They also gave way when the soldier and robot walked past. The soldier's presence could additionally trigger civilian reactions such as deference or anger. The soldier's objectives were to look for weapons, contraband, or potential threats. The ADAPT Behavior Trees scripted all interactive NPC behaviors.

ADAPT's strength was also its limitation. Since actions were scripted, one had to write programs to manage all contingencies. In the real world, such determinism is awkward or impossible. Consider a computer program to play chess. Theoretically, one could create a branch for each possible move, but the number of branches rather quickly exceeds any known storage medium. Chess-playing programs instead work with strategies and planning to achieve successful outcomes. So, too, did we realize that what ADAPT needed was an AI planning capability. Agents could have goals that were states and not just locations. Shoulson and Kapadia dove into story scripting using planning [3]. Shoulson took a semester sabbatical from his Ph.D. to visit Disney Research Studios in Zurich,

Switzerland. There he finalized his narrative planning system "CANVAS" and constructed animated interactive examples around a jailbreak and a bank heist [4]. Shoulson completed his Ph.D. in 2015 and, unsurprisingly, joined the game industry.

Kapadia had other interests in addition to storytelling. Navigation in complex environments was his forté. Although he kindly placed my name on many of these publications since he was my postdoc, the ideas were his. He and Nuria Pelechano led a team that included her Ph.D. student, Alejandro Porres (I love these multi-generational teams!), CGGT Master's student Vivek Reddy, and visiting student Francisco Garcia in developing methods for multi-resolution path planning [5]. Multi-resolution paths examined local, medium, and long-range geometry to compute an optimal path to a goal, even in the presence of moving obstacles. Kapadia and Garcia also collaborated on constraint-aware navigation with Shoulson and CGGT Master's student Kai Ninomiya [6]. Constraint-aware navigation uses semantic information on terrain types, for example, grass, water, or roads, to obey rules and influence optimal path selection. It could also use line-of-sight information to have an agent choose a path to avoid being seen by other agents or sensors, a handy skill in first-person shooter games.

Some of Kapadia's earlier work at UCLA involved a synthetic perceptual sense to enable agents to better navigate a crowded scene. He built a testbed for evaluating agent navigation in crowds [7]. He speculated that automated methods might be helpful in formulating an agent's movement strategy rather than relying on manually tuned parameters. As the age of AI and machine learning methods began to take over many of the more classical representations in computer graphics, we gave this topic to another of my Ph.D. students, Cory Boatright. Working with DMD undergraduate Jenny Shapira, Boatright successfully produced synthetic crowd motion data usable by machine learning procedures [8]. Synthetic data had several advantages over existing alternatives: real empirical crowd data or manually generated test cases. The former was sparse and highly dependent on geometric context and available data resources; the latter was fraught with bias and tedium. Although remarkably nervous at his Ph.D. defense in 2015, he did well and became the first person in his family to achieve that honor.

References

1. D. Slonneger, M. Croop, J. Cytryn, J. T. Kider Jr., R. Rabbitz, E. Halpern and N. Badler. "Human model reaching, grasping, looking and sitting using smart objects." Proc. International Ergonomic Association—Digital Human Modeling, Lyon, France, 2011.
2. A. Shoulson, N. Marshak, M. Kapadia and N. Badler. "ADAPT: The Agent Development and Prototyping Testbed." IEEE Transactions on Visualization and Computer Graphics (*TVCG*) 20(7), July 2014, pp. 1035–1047.
3. A. Shoulson, M. Kapadia and N. Badler. "Toward event-centric interactive narrative." Intelligent Narrative Technologies 6, 2013.
4. M. Kapadia, S. Frey, A. Shoulson, R. Sumner and M. Gross. "CANVAS: Computer-Assisted Narrative Animation Synthesis. Symposium on Computer Animation 2016, pp. 199–209.

5. M. Kapadia, A. Porres, F. Garcia, V. Reddy, N. Pelechano and N. Badler. "Multi-domain real-time planning in dynamic environments." ACM SIGGRAPH/Eurographics Symposium on Computer Animation (SCA) 2013.
6. M. Kapadia, K. Ninomiya, A. Shoulson, F. Garcia and N. Badler. "Planning approaches to constraint-aware navigation in dynamic environments." Computer Animation and Virtual Worlds, Vol. 26(2), March/April 2015, pp. 119–139.
7. S. Singh, M. Kapadia, P. Faloutsos and G. Reinman. "Steerbench: a benchmark suite for evaluating steering behaviors." Computer Animation and Virtual Worlds, Vol. 20 (5–6), 2009, pp. 533–548.
8. C. Boatright, M. Kapadia, J. Shapira and N. Badler. "Generating a multiplicity of policies for agent steering in crowd simulation." Computer Animation and Virtual Worlds, 2014.

Smells and Sounds

38

The evolution of digital humans took obvious paths toward reproducing enough body structures to enable crucial musculoskeletal behaviors such as reaching, grasping, walking, and looking. Concurrently, many other researchers explored the conditions and requirements for such creations to possess the appearance of intelligence. "Artificial Intelligence" studies mechanisms for logical reasoning, action planning, language generation, and speech production. From early in computer science, the study of how humans might process visual input fascinated computer vision and robotics researchers. Computer graphics human models naturally adopted some visual sensing capabilities, mostly for navigating around obstacles and avoiding collisions with other agents. Artificial vision capabilities became important for real-time virtual human animations as early as the 1990s [1]. We explored visual attention properties, too [2]. Sound effects appeared in computer games. There was little incentive to provide agents with hearing capabilities because, in a computational environment, one could simply pass digital data messages directly from a sound source to an agent. A sound could possess a radius of influence, perhaps, so that only nearby agents would receive the sound information. Other human senses of taste and smell seemed, at first, less relevant to virtual human existence.

As on so many other occasions in my life, a chance meeting triggered a new direction for investigation. Likely at some ergonomics conference, I met someone from the U.S. Bureau of Mines. We talked about mine safety: mining being a particularly hazardous occupation. Mines are a different kind of workplace than manufacturing plants and vehicle cockpits. He hinted that funds might be available to put *Jack* to work in a mine so we could explore these safety issues. Since by that time in the late 2000's *Jack* was already a commercial product, we thought that we might add suitable higher-level behaviors for mine applications. One addition would be evacuation during emergencies, something we

N. Badler, *On Raising a Digital Human*, Synthesis Lectures on Computer Science, https://doi.org/10.1007/978-3-031-63945-6_38

were already studying with our crowd simulation systems. The interesting part, however, turned out to be the *sensing* that something was wrong and that evacuation was necessary.

Rather coincidentally, we had a lab visitor from Brunel University in the UK, Dr. Jinsheng Kang, who was interested in the human response to airborne pollutants, such as smoke. The Bureau of Mines connection supplied us with some mine layout maps. In a pattern that had become the norm for post-2000 projects in the HMS Center, we assembled a diverse group to undertake a mine simulation. Besides Dr. Kang, we engaged three Ph.D. students, Pengfei Huang, Joe Kider, and Ben Sunshine-Hill, a DMD undergraduate, Jon McCaffrey, and a visiting student from the University of Puerto Rico, Desiree Velázquez Rios. They modeled the mine, mining equipment and, most notably, the air flows through the mine passageways.

We weren't interested in the normal everyday workings of a mining crew. Rather, we looked at what happened in an emergency [3]. The public generally hears of mine collapses, but more insidious are mine fires and the noxious gasses they generate, such as carbon monoxide and hydrogen chloride. We constructed a physiological exposure model based on human tolerance to these toxins. Our MediSim experience proved valuable here. We also adopted the communication model from Pelechano's HiDAC, so that agents could inform others of fires and known path blockages. When a fire broke out, any nearby agents would abandon their machines and run toward the nearest exit. They informed other agents they encountered and exchanged any information on blocked exit routes. Meanwhile, the noxious fumes from the fire circulated in the mine passageways according to its tunnel and ventilation structure. A visible fog effectively portrayed the smoke and fumes in the graphics animation. While we can't really call it a "sense of smell," each agent proximal to these airborne substances accumulated potentially unhealthful exposures by inhalation. If these exposure levels exceeded a first threshold, they stopped running and dropped. If the exposure exceeded a second, higher threshold, they died. Although rather depressing, the simulation effectively communicated the seriousness of environmental dangers. We were unsuccessful in obtaining any subsequent funding after this pilot project, but it was exciting to watch—in a macabre sort of way—as a disaster unfolded.

Around the contemporary backdrop of smart agents and VR experiences with animated agents, I realized that we had neglected the role of human audition in creating a simulated world. The canapé party VR demonstration led me down the rabbit hole of the "cocktail party effect": in a room full of overlapping conversations, a listener could selectively focus attention on one to pick out distinct utterances. I did not initially possess sufficient signal processing skills to follow this route mathematically, but the property itself piqued my curiosity enough to try to emulate the effect, if not the actual mechanism.

Pengfei Huang, motivated by the VR cocktail party crowd simulation and his MIMOSA mine simulation, turned his attention to auditory signal processing. Huang, Kapadia, and I decomposed sound signal processing into five major components: generation, propagation, attenuation, degradation, and perception. We named this system SPREAD [4]. To keep simulations running in real time, we assumed that all sound sources and hearing agents would be at approximately the same height (2 m) on a 2D plane. Sound signal amplitudes were stored in twenty frequency bins across the audible spectrum with temporal samples covering a one-second duration. A preprocessed dataset of 100 sounds formed the database for the generation step [5]. A number of sound sources could be located anywhere in the 2D plane. Each source could be assigned a temporal sequence of sounds. The sources could be fixed or moving. Animated *Jack* agents themselves could be sound sources as well as sound absorbers.

Sound propagates through an environment in expanding waves, much like the circular expanding ripples in water caused by a dropped stone. Sounds can be absorbed by objects (including people), reflected from flat surfaces, refracted (bent around objects), and attenuated (lose power) as they become distant from the source. These modifiers depend on the original amplitude (loud sounds travel farther), the material composition (absorption) of environmental surfaces, and the frequency distribution of the sound itself (high frequencies decay faster). Huang simulated all these propagation effects with a 2D, cellular, iterative, "transmission line" algorithm. Not only did sound amplitude decrease during propagation but the frequency profile of a sound also changed. In practice, this meant that the sound that reached a listening agent's virtual "ear" was not just a quieter version of the original: it may have lost high frequencies and developed delays and echoes from bouncing around obstacles.

Each agent required a sound perception process to attempt to categorize any incoming sound and localize its probable direction [6]. A published psychophysical study [5] of sound perception greatly aided this perceptual component. A hierarchical tree of sound types contributed a novel feature of the perception process. A sound might arrive at the agent without much distortion, in which case the sound perceived was that best matched by one of the database sound samples, for example, "banging." However, with distortions, the best match might be a more general category, such as "destructive sound," or even more general, a "single impact." The root of the tree was just "some sound," if it was otherwise unidentifiable. Of course, with sufficient attenuation or distortion, a sound might not be perceived at all.

One of the fun demos we produced featured a "drill sergeant" with a whistle. The marching "troops" had to turn to orient toward the whistle, provided that they could hear and identify it. Sometimes the folks in the back could not clearly perceive the whistle due to attenuation from the other people in front of them. Since they lost the signal, they failed to orient toward it until they could again hear it clearly. For another dramatic example, we created an audio-driven version of the Atari game "Frogger." Our virtual—and blind—frogs had to hear the location and estimate the velocity of oncoming traffic to hop successfully from lane to lane while crossing the road. That was a nail-biter, but it worked!

Since speech phonemes were also propagated and distorted, agent-to-agent communications via speech might not be correctly perceived if the inter-agent distance is too great or if environmental objects interfered. Although we began to explore the language understanding degradation process, we did not have sufficient time nor energy to report on it. Had we been able to so, we might have attained some insights into, or at least a demonstration of, the cocktail party effect.

We believed, however, that SPREAD had both utility and novelty. We approached the Penn Center for Innovation and asked if they would consider it sufficient for a U.S. patent application. They agreed. On April 10, 2018, Huang, Kapadia, and I received patent number 9,942,683: "Sound Propagation and Perception for Autonomous Agents in Dynamic Environments." Huang had since moved to Microsoft and Kapadia joined the computer science faculty at Rutgers University. The patent was a nice capstone to Kapadia's term as my postdoc.

References

1. O. Renault, N. Magnenat-Thalmann and D. Thalmann. "A vision-based approach to behavioral animation." Visualization and Computer Animation, Vol. 1(1), 1990, pp. 18–21.
2. S. Chopra-Khullar and N. Badler. "Where to look? Automating attending behaviors of virtual human characters." Autonomous Agents and Multi-agent Systems 4(1/2), 2001, pp. 9–23.
3. P. Huang, J. Kang, J. Kider Jr., B. Sunshine-Hill, J. McCaffrey, D. Velázquez Rios and N. Badler. "Real-time evacuation simulation in mine interior model of smoke and action." Computer Animation and Social Agents (CASA) 2010.
4. P. Huang, M. Kapadia and N. Badler. "SPREAD: Sound propagation and perception for autonomous agents in dynamic environments." ACM SIGGRAPH/Eurographics Symposium on Computer Animation (SCA) 2013.
5. B. Gygi, G. Kidd and C. Watson. "Similarity and categorization of environmental sounds." Attention, Perception, and Psychophysics 69(6), 2007, pp. 839–855.
6. Y. Wang, M. Kapadia, P. Huang, L. Kavan and N. Badler. "Sound localization and multi-modal steering for autonomous virtual agents." Proc. I3D, 2014.

Running Hot and Cold

I was never a big fan of using the telephone. I could write letters, but the mail was slow. When the technology became painless, email was my communication medium of choice. My stints as department Chair and then Associate Dean ramped up my email loads to hundreds per day. I often felt like I was just a router, forwarding emails along to more appropriately responsible individuals. However, I also liked the mystery or surprise factor in emails: maybe they came from someone interesting, a prospective Ph.D. student, or a potential international visitor.

So, it was exciting to receive an email from Soraia Raupp Musse asking me whether she and her husband Claudio Rosito Jung could spend their academic year sabbatical in Philadelphia. I knew Musse well from her crowd simulation work with Daniel Thalmann, having met her many years before in Switzerland. She had also sent us her excellent Brazilian students, Rossana Baptista Queiroz and Vini Cassol, for their Ph.D. "sandwich" years. The plan would be for Musse to spend 2015–2016 in my lab and for Jung, a computer vision researcher, to visit in the Penn GRASP Lab. They would bring along their two children and rent a house for the year. As plans go, it was a good plan. However, for reasons still unknown, GRASP never welcomed Jung to have a desk in their lab space. I gladly gave him a workstation in HMS.

A few years earlier, in 2013, my architecture colleague Ali Malkawi hosted a "Building Simulation World Summit" on the Penn campus. I talked about Nuria Pelechano's crowd simulation work and Jan Allbeck's CAROSA building occupant simulation. One of the invited speakers, Vladimir Bazjanac from Stanford, talked about architectural concepts I had never considered: human *comfort*. Of course! Buildings are not just used for ingress and egress; people *work* there. Moreover, emergency evacuation studies are contingencies one hopes don't actually occur. The people working inside need to be comfortable

© The Author(s), under exclusive license to Springer Nature Switzerland AG 2025   181
N. Badler, *On Raising a Digital Human*, Synthesis Lectures on Computer Science,
https://doi.org/10.1007/978-3-031-63945-6_39

all the time! Heating and air conditioning engineers know this domain well. There are international standards for assessing human comfort as a function of temperature, humidity, airflow, work activity, and clothing [1]. This was a revelation, as consideration of non-strength comfort never arose in the *Jack* environments. For the next year or two, Bazjanac, his colleague Renate Fruchter, and I tried to figure out how to obtain funding to combine 3D building models with human comfort models. I had no spare, unoccupied Ph.D. students. I would need funding to bring in someone new to concentrate on this connection.

What we could do, relatively inexpensively, is work with undergraduate students during the summer break. In anticipation of having Musse and Jung visit, I organized a 2015 summer project to create an interactive 3D model of a famous Philadelphia destination, the Reading Terminal Market. This is an extensive downtown market with numerous individual vendors, ample seating, restaurants, and boutique gift shops. It occupies a full city block on one floor, with exterior door access. Lab alumni couple Dawn and Tripp Becket, along with Diane Chi, helped fund the participating DMD, CGGT, and computer science students: Adam Mally, Charles Wang, Kenji Endo, Mike Rabbitz, Nihaar Narayan, and Caroline Smith. They photographed, measured, and modeled the entire market environment in Maya, including vendor stalls, seating, and architectural structures, then imported and lit the model in the Unity game engine. A NavMesh graph structure represented the walkable environment for ADAPT human agents. Automatic doors opened and closed as an agent approached an exit, just as in reality.

Jung and Musse settled into HMS in September 2015. Late summer can still be very hot in Philadelphia. I was outside and noticed a group of people, perhaps waiting for a bus, crowded underneath a canopy to escape from the direct sun. There was plenty of room outside the canopy, but shade was apparently more important to them than space. I think I even took a picture of the group. I showed this to Musse and Jung and wondered whether heat could be an unexplored factor in crowd behaviors. We couldn't find any overt references to temperature influences on crowd motions other than the obvious desire to escape a fire by running away from it. Even our simulated mine workers did that!

The idea stuck. Jung gladly took on the lead role and recruited another Ph.D. student visitor already present in our lab, Ms. Lu Chen, from Ocean University in China. Chen and two other Penn DMD undergraduates, Mohamad Moneimne and Charles Wang, did the programming work. They situated the human agents in the Reading Terminal Market model. This space had hot locations, such as pizza ovens, and potentially cold spaces, such as the external entrance doors. The agents themselves were local heat sources. The market model would be ideal for simultaneously exploring crowd movements in response to crowd density and to sensed and changing temperature conditions.

Jung programmed a thermal propagation model for any 2D floor plan. The NavMesh could also contain the local temperature values for any walkable space, so every agent was aware of their immediate thermal environment. In general, agents preferred to be comfortable relative to their sensed temperature and the density of agents immediately around them [2]. An agent experiencing too warm conditions might first remove their outerwear, then move away from a local heat source, and try to reduce their proximity to neighboring agents. Conversely, an agent feeling too cold might don their jacket, try to move to a warmer spot, or crowd closer to other agents to conserve body heat. One surprisingly delightful consequence of this is that agents sometimes moved because of thermal conditions and not because they had a destination goal to achieve. For example, agents standing near an exterior door on a very cold day might spontaneously move away from the door if it opened and let in cold air. In other crowd simulations, a lack of a distinct goal meant a lack of movement. Our agents could spontaneously move to a more comfortable location of their own choosing due to environmental considerations alone.

Thermal comfort led to another unexpected insight: people do not make comfort decisions instantaneously. Someone might just wait out a brief and transient temperature change. One might go outside without a coat on a cold day if the expectation is that the exposure will be brief. Likewise, one can tolerate walking past a hot pizza oven but probably wouldn't linger next to it having a long conversation with a friend. These observations led us to have each agent *integrate* its local thermal conditions over an exposure time "sliding window" of approximately 10 s. We chose 10 s to keep the simulation interesting, but there's otherwise nothing special about this duration. We exploited a mathematical concept called *hysteresis*: the tendency of a two-state system to stay in its current state even in the presence of disruptors. The state may flip when a triggering event occurs, a threshold is exceeded, or a duration elapses. Notably, the triggering condition may differ depending on the direction of the state change.

Overall, this project changed our perspectives on non-emergency crowd behaviors in three distinct areas. First, rather than "one agent, one goal" directives, individual agents made *non-spatial* goal-directed movement decisions. Second, thermal and crowd density stimuli were integrated over a temporal window to portray hysteresis and avoid instant responses. And, third, we learned that humans do not all respond to comfort in the same way. Someone may like it warmer; someone else cooler. Individuals may have personal or cultural biases on what constitutes "closeness" to neighboring agents. To model these variations, we could have explicitly programmed thresholds and preferences into each agent, but that would have been tedious and arbitrary. Instead, we took our cue from the international comfort standards and made our agents respond *probabilistically* [1]. Having agents behave within probabilistic distributions was not in itself a new idea. Rather standard procedures for reducing crowd monotony existed: agents could be assigned velocities within a reasonable walking speed range, their gaits might be distinctive [3], or clothing or carried items could vary [4]. We had just not considered comfort responses among the range of influences.

While the Reading Terminal Market simulations were cute, I started to think about more serious applications. Real-world problems informed our *Jack* developments, and I felt that I had lost those sorts of tangible connections. I began to look into descriptions of major non-conflict crowd disasters. These fell into a few categories, including stampedes and crushes. Stampedes cause casualties when people lose their footing. They have no way of getting up due to crowd density and movement. Crowd pressure against an immovable obstacle such as a wall or barrier causes high-density packing crushes. People literally have no room to breathe. As Musse, Jung, and I thought about these further, we wondered if there might be another causal condition for crowd casualties: thermal overloading. Sure enough, there had been a recent large-scale tragedy at the Hajj in Mina, Mecca, Saudi Arabia, on September 24, 2015, where more than 2000 pilgrims lost their lives [5]. Despite being called both a crush and a stampede, we nonetheless felt that intolerable thermal loads under high-density crowding could have been a causal factor. We abstracted the geometric situation into two intersecting streets and modeled 11,000 individuals merging toward their common goal. The outside temperature was likely around 95 °F in bright sunshine. Even by conservative estimates of crowd density, and based on the international comfort standard, numerous individuals would have registered a body discomfort index greatly exceeding human tolerance levels. Under our analysis, many of the victims may have fallen, fainted, or stopped breathing from thermal stress, rather than from overt pressure crushes. We will never know for sure, but the simulation, in our opinion anyway, is compelling.

After Musse and Jung returned home to Brazil, we kept in touch through their student Vini Cassol's Ph.D. work. He based some of his crowd evacuation models on another tragedy local to their community, the Brazil nightclub fire in Rio Grande do Sul on January 27, 2013, which killed 242 people [6]. Although the nightclub's only open exit clearly contributed to the high casualty rate, Cassol also hypothesized that at least some of the occupants may have been disoriented or even unaware of the dire situation due to *intoxication*. His observation added another physiological influence into human crowd behavior [7].

I turned once more to the physiology of crowd disasters. In 2017, we hosted a visiting faculty member from Southeast University in China, Dr. Libo Sun. Sun had already spent a year in HMS when she was a Ph.D. student and worked on a conversation simulation project [8]. She wanted to engage in some new research to boost her faculty career, so I welcomed her return visit. We cast about for new ideas until we hit on the crowd-crush situation. It's worth noting that many computer graphics crowd simulations are generous with the amount of space a virtual person occupies, to create pedestrians without interpenetrating bodies and limbs. Even a dense crowd often looks like individuals are respectfully and visually separated. High-density crowd simulations often simplify graphics display issues by abstracting bodies to compact cylinders (essentially ignoring swinging arms) or assuming that occasional arm-to-arm collisions will not be noticeable in a dense crowd. Note, too, that dense crowds often move slowly and do not leave much room anyway for normal gait arm swinging.

Against this background, Sun and I wondered what would happen if we placed complete human body models (with arms and legs) into a physics-based simulation and submitted them to crushing forces. That is, suppose they were packed into a space where they could not move or escape, yet the density steadily increased. Our simulation consisted of a group of individuals in a fixed three-sided space who were compressed by a fourth wall driven by a piston. The piston represented the generalized crushing force of other crowd members. The Havoc engine in Unity supplied the physical simulation. Sure enough, we were able to show that numerous individuals experienced densities exceeding human tolerance levels and would be casualties under such conditions [9]. *Post facto* simulations cannot change what happened in a disaster. Our hope is that simulations may help identify and mitigate other potentially dangerous spatial situations.

As an academic, we often judge the quality or impact of a publication by the number of citations it subsequently receives. That measure does not include Internet search hits by the curious. I was surprised when a South Korean doctor contacted me shortly after the October 29, 2022, Seoul Itaewon crowd tragedy. He wanted to study the disaster to help avert any reoccurrence. I didn't feel I had much to offer other than what we learned from Sun's crowd crush simulations, but I remotely attended a couple of meetings. Soon, though, I bowed out of the discussion. Commercial software systems exist to analyze and study real-life crowd movements. Moreover, there is a robust community of evacuation simulation experts. They are better prepared to address influences from urban design, transportation access, police presence, and local political structures. Computer graphics can provide nice visualizations but cannot by itself readily address such social issues.

References

1. *ASHRAE Standard 55–2004: Thermal environmental conditions for human occupancy.* American Society of Heating, Refrigerating and Air-Conditioning Engineers, Atlanta, GA, 2004.
2. L. Chen, C.R. Jung, S.R. Musse, M. Moneimne, C. Wang, R. Fruchter, V. Bazjanac, G. Chen and N.I. Badler. "Crowd simulation incorporating thermal environments and responsive behaviors." Presence J., 26(4), Fall 2017, pp. 436–452.
3. K. Ashida, S.-J. Lee, J. Allbeck, H. Sun, N. Badler, and D. Metaxas. "Pedestrians: Creating agent behaviors through statistical analysis of observation data." Proc. Computer Animation November 2001, Seoul, South Korea, pp. 84–92.
4. E. Wolf and N. Badler. "The distribution of carried items in urban environments." Presence: Teleoperators and Virtual Environments J., MIT Press, Vol. 24(3), Summer 2015, pp. 187–200.
5. https://en.wikipedia.org/wiki/2015_Mina_stampede; accessed November 7, 2023.
6. https://en.wikipedia.org/wiki/Kiss_nightclub_fire; accessed November 7, 2023.
7. V. Cassol, S. Musse, C. Jung and N. Badler. *Simulating Crowds in Egress Scenarios.* Springer, 2018.
8. L. Sun, A. Shoulson, P. Huang, N. Nelson, W. Qin, A. Nenkova and N. Badler. "Animating synthetic dyadic conversations with variations based on context and agent attributes." Computer Animation and Virtual Worlds J., 2012.
9. L. Sun and N. Badler. "Exploring the consequences of crowd compression through physics-based simulation." Sensors 18(12), 2018, pp. 4149.

Over more than 40 years of investigating human movement, I came to realize a few truths. I certainly don't claim to have discovered them, but they became clear motivating factors for our own studies. First, human movement isn't random. Second, what we perceive may not have much to do with the way motions are internally generated, say by muscle activations. Third, movement is motivated by innate psychological forces or characteristics: perhaps mood, emotion, or personality. Fourth, nonverbal movements alone can convey aspects of the internal human state. Let's look at each of these.

Human movement is not random. In the early days of animating human models, random arm movements were supposed to project animacy, or at least dispel the appearance of body rigidity. I found these movements more disconcerting than convincing, because they had no correlation to the communicative message in speech, for example, nor did they acknowledge the speaker or the shared environment. Justine Cassell alerted us to the need for gesture compatibility, while our work on *Jack* Presenter engaged us in generating contextual gestures referencing the environment. Cartoon animators long recognized that a lack of character movement broke the illusion of animacy. The movement had to be the viewer's window into the character's attitudes, thoughts, and intentions. Random movement projected the appearance of random behavior.

This leads to the second truth: whatever the mechanism might be that generates character motion, our perception and interpretation of that motion is, literally, what counts. Should we have built a detailed human musculature model? If we had, we still would have no idea what the control signals might be to achieve a given communicative purpose. Moreover, muscles vary across individuals (for example, most of us are not Olympic athletes), and the control patterns are likely learned over years of repetition and coordination. Another perspective on motion appearance is that skilled actors can convey their internal emotional state even if that is not their "normal" behavior. This takes practice, of course, but it is possible. What ultimately matters is whether the character portrayal conveys the proper attitudes, thoughts, and intentions and, moreover, that they are *consistent* across all body communicative channels. It is not surprising that this is the same requirement as an effective cartoon. More evidence that these communicative purposes arise from motion alone is that an animated character's face and body movements may be sourced directly from motion capture. The movements of live actors transform into numerically expressed joint angles, skin deformations, and facial expressions. Algorithmic musculoskeletal models need not drive those physiological systems, although facial animation comes closest. Reproduction accuracy brings us to the brink of the uncanny valley; hence, modified, exaggerated, or caricatured 3D models might avoid unwelcome attention to artificiality. Our own human models never achieved a detailed appearance close enough to actual human shape such that the uncanny valley interfered with their applications or use cases. In fact, we wanted to make the model's motions look convincing, rather than its shape details. Even though *Jack*'s "turtle" torso shape segments looked rather unrealistic, the fact that the bending and twisting motions were accurate for a human vertebral spine made appearances acceptable and complex working movements possible.

Third, human needs, physical behaviors, and psychological factors motivate movements. Virtual human actors should breathe and, if fatigued, exhibit corresponding exaggerated chest movements [1]. *Jack* and other human models reproduced physical behaviors such as reaching, grasping, walking, and balancing. When these actions have communicative intent rather than just goal achievement, the problem becomes more difficult. For example, how one reaches for an object says a lot about the person's attitude toward it: is it a rapid snatch that communicates aggression and a need for possession, or is it a delicate and tentative pluck because it is fragile, hot, or revered? These distinctions motivated our study of Laban Movement Analysis. The *how* of a movement mattered more than the *what*.

Lastly, someone introduced us to Ambadi and Rosenthal's seminal work on human personality and nonverbal communication [2]. By showing subjects short videos of teachers *without any audio*, the subjects were able to judge teaching effectiveness consonant with prior student assessments. Their initial video samples were only 10 s long, but they also reproduced similar results with even shorter time exposures. Ambadi and Rosenthal showed that subjects were not basing judgments on verbal content or attractiveness. The implication is that "thin slices" of nonverbal body and gestural movement alone are sufficient to create valid evidence of personality. Indeed, first impressions matter.

These truths guided our quest to understand how to turn LMA concepts into animation procedures. Although I started my LMA analyses by believing I needed physics simulations, Diane Chi's EMOTE demonstrated that well-constructed kinematics techniques were apparently adequate. We realized that using LMA or its EMOTE implementation was not just about gesture: consistent expressiveness pervaded the entire body animation.

Like the old schoolyard trick of trying to pat your head while rubbing your stomach, LMA expressiveness could not (easily) be varied across body units. A real actor would have difficulty decoupling LMA characteristics in one arm for a different set in the other. These couplings extended beyond the arms to head, face, torso, and leg motions as well [3]. One motivation for building a digital human model driven by LMA factors was to create psychological experiment datasets with deliberate mismatches to determine how readily subjects could perceive whether "something seemed wrong" in an animated character.

We were convinced that internal psychological motivations affected external movements. LMA itself, however, was agnostic to internal state. LMA, as a tool used to analyze movement content, deliberately avoided emotion-laden terms. In fact, not every movement had to exhibit LMA factors. One curious expression used by LMA notators was that LMA characteristics "crystallized" out of the movement stream. There was no mixed metaphor here: many movements did not or had no reason to portray communicative intent; they were "just" movements. When movements demonstrated some detectable LMA factors, they became noteworthy and perhaps significant as an insight into the internal mental state of the performer. Modern dance, mime, and ballet thrive on nonverbal movement as a medium of expression where performers must convey emotions, intentions, and attitudes to an audience. But what connection did LMA have to internal psychological state? That was a question we could not answer until the mid-2010s.

Unexpectedly, crowd simulation work helped us address this question. In her Ph.D. thesis from Bilkent University in Ankara, Turkey, Funda Durupinar examined the movement characteristics that distinguished the visual appearance of different crowd types, such as mobs, parades, or pedestrians [4]. After completing her degree, Durupinar moved to the US. While I still had some research funds, I hired her as a postdoc. If a crowd could have an emotional state or a personality, then surely an analogous approach could portray the same in an individual.

More than 15 years after our original EMOTE work, we would have to revisit LMA and determine how it could be related to personality. We chose personality rather than emotion because it is a relatively stable characteristic of an individual. Emotions and moods are shorter duration states that are heavily influenced by external context, events, or other people. Durupinar selected the common and respected five-factor OCEAN model as the personality descriptor: Openness, Conscientiousness, Extroversion, Agreeableness, and Neuroticism. The overall goal was simple to state: provide a virtual human model with a selected OCEAN personality, give them a movement to perform, and have them do the motion influenced only by their personality. Change the personality, and the performance will change. Change the underlying motion, and a consistent personality impression will still come through.

We needed to establish a computational mapping from the OCEAN model to animation parameters. It could have been tempting to throw a machine learning solution at this, but we did not have any large dataset of annotated motions or live subjects to cover all the OCEAN combinations. Our approach split the OCEAN to animation mapping into two mappings: one from OCEAN into LMA factors, and a second from LMA to animation parameters. EMOTE itself was adequate proof that we could do the latter, so we needed to focus on the former. We recruited the expertise of a dance instructor who taught LMA at neighboring Drexel University, Professor Susan Deutsch. I also solicited contributions from a University of California at Davis colleague in computer graphics, Michael Neff. When he defended his computer science Ph.D. thesis from the University of Toronto, I was his external reader. Neff had studied LMA, implemented his own animation systems there, and had recently earned an LMA notator certificate. I provided the environment and the incentive, but the others did all the work. Durupinar named the new project PERFORM: Perceptual Approach for Adding OCEAN Personality to Human Motion Using Laban Movement Analysis [5]. Note the emphasis on "perception" as the basis for evaluating our success. She implemented PERFORM in the Unity game engine for the animations.

First, we needed to revisit the LMA-to-EMOTE mapping. Durupinar designed a simple user interface with sliders to adjust all the EMOTE parameters. Deutsch and her students would use the sliders to iteratively manipulate and animate a human model to produce motions that conformed to notator beliefs. People rarely exhibit one LMA Effort factor in isolation. In LMA parlance, the co-occurrence of two extreme efforts is called a *State*, and three together is a *Drive*. States are common and thus not diagnostic for a specific personality type. Drives, however, are considered extraordinary motions and are indicative of intense feelings; hence, they may be readily identifiable as salient indicators of personality [6]. Therefore, we chose to animate the 32 possible Drives to derive the LMA-to-animation mapping. The expert notators observed a neutral gendered 3D wooden mannequin model while they adjusted the low-level animation parameters to produce a satisfactory animated performance for each Drive. This updated EMOTE to give us a new LMA-to-animation mapping grounded in expert knowledge and motion perception.

The major gap, and hypothesis, lay in creating an OCEAN-to-LMA mapping. There was no body of data to inform the creation of such a mapping or even provide evidence of its existence. We only knew that skilled actors or dancers could portray personality attributes through movement. The five OCEAN factors would set the desired personality of an animated figure, but how the OCEAN factors were realized in LMA terms would be tricky. We made two major decisions. The first would be to use a cross-section anonymous subject pool obtained through Amazon Mechanical Turk. The second involved how these subjects would identify a personality type or an LMA drive in an animation when the probability of them knowing anything about either OCEAN or LMA would be essentially zero. The key turned out to be the ten-question TIPI test, which uses simple English sentences to provide a validated OCEAN assessment of personality type [7]. Subjects watched two animated characters side-by-side on the screen and answered questions such as "Which character looks MORE *open to new experiences & complex* and LESS *conventional & uncreative?*" The salient words came directly from the TIPI questions and were diagnostic for one of the OCEAN dimensions. Subjects could answer *left, right,* or *neither*; they did not need to know anything about LMA, as the experiment selected and animated the figures based on the chosen drive. Fortunately, we registered good correlations between the identified personality types and the motion's performances through the LMA middle layer. A final experiment using Amazon Mechanical Turk subjects validated the veracity of the full personality-to-motion transformation.

Our team had achieved one of my long-term goals: to realize individual personality expression through movement alone. The capability was surprisingly simple. Setting five sliders on the OCEAN dimensions specified a character's personality. The selected OCEAN-type preset internal LMA factors which were established by the PERFORM experiments. The character's movement could be selected from any source, including motion capture, procedural generation, or manual animation. When played through the LMA presets, the character portrayed its personality through the modified movements. We could animate a crowd who ostensibly performed the same basic motion but who nonetheless did so with individualized personality-based nuances. We took motion capture from online databases and modified disco dances, ball playing, and spectator cheers.

Our prior Army human-robot marketplace simulation triggered a thought: if we were to simulate a populace visually situated in some part of the world, should we model cultural cues to their behavior? In other words, should we be concerned about portraying stereotypes for good (to look "appropriate") or for bad (because they are stereotypes)? Durupinar and I teamed up with another Penn CIS colleague in natural language processing, Professor Ani Nenkova, and a CGGT Master's student, Kuan Wang. Nenkova pointed us to two interesting word lists of 135 nationalities and 98 professions. People often have stereotypical verbal impressions of some of these terms [8]. Did these stereotypes extend to visible personality traits exemplified by otherwise neutral motions, such as walking? We used PERFORM to create opposed pairs of OCEAN traits. Amazon Mechanical Turk subjects (all based in the US, to minimize cross-cultural bias) judged whether these simultaneously presented animation pairs appeared more like their own conception of one culture or another. We ran analogous experiments for the profession categories. There were some clear OCEAN personality biases among both cultures and professions. When we turned it around and asked if an animation resembled either of the cultural groups who had appeared at opposite ends of a specific OCEAN dimension, the matchings were no better than chance. However, they were significant for many of the profession types. In a way, this result was heartening: cultural stereotypes were seemingly unimportant to animation, although personality types associated with specific occupations did matter.

Kuan Wang used the Unity game engine to construct a high-level character interface to PERFORM. A game designer could select a human agent and designate their culture and profession. Wang's interface displayed the OCEAN factors from both sources, which could be weighted and blended together or even set individually [9]. These OCEAN personality factors then biased the character's animation through the PERFORM model. We imagined some novel applications for this interface, such as customizing a game to a particular genre or taking an existing game and debiasing character motions to dispel cultural stereotypes.

For me, PERFORM closed a longstanding research gap. We had been able to demonstrate convincingly that an OCEAN personality type could affect an arbitrary motion performance in a quantifiable fashion, and did so through a route—LMA—led by the movement notation community. Of course, as research, there were other consequences to be explored, but I was quite content to leave those to my colleagues Durupinar, Neff, and Nenkova. The PERFORM publication has garnered enough citations to leave us with a strong impression of its significance, and hopeful impact, on the virtual human animation community. I have enjoyed the sense of closure that comes from wrapping a big problem in an understandable box and topping it with a bow. I can give it away.

References

1. J. Kider Jr., K. Pollock and A. Safonova. "A data-driven appearance model for human fatigue." Eurographics/ACM SIGGRAPH Symposium on Computer Animation, pp. 119–128, 2011.
2. N. Ambady and R. Rosenthal. "Thin slices of expressive behavior as predictors of interpersonal consequences: A meta-analysis." Psychological Bulletin, 111(2), 1992, pp. 256–274.
3. N. Badler and J. Allbeck. "Towards behavioral consistency in animated agents." In *Deformable Avatars*, N. Magnenat-Thalmann and D. Thalmann (Eds.), Kluwer Academic Publishers, 2001, pp. 191–205.
4. F. Durupinar, U. Gudukbay, A. Aman and N. Badler. "Psychological parameters for crowd simulation: From audiences to mobs." IEEE Transactions on Visualization and Computer Graphics, Vol. 22(9), 2016, pp. 2145–2159.
5. F. Durupinar, M. Kapadia, S. Deutsch, M. Neff and N. Badler. "PERFORM: Perceptual Approach for Adding OCEAN Personality to Human Motion using Laban Movement Analysis." ACM Transactions on Graphics, 2017.
6. B. Adrian. *An Introduction to LMA for Actors: A Historical, Theoretical, and Practical Perspective.* Allworth Press, New York, pp. 73–84, 2002.
7. S. Gosling, P. Rentfrow and W. Swann. "A very brief measure of the big-five personality domains." Journal of Research in Personality, 37(6), 2003, pp. 504–528.
8. O. Agarwal, F. Durupinar, N. Badler and A. Nenkova. "Word embeddings (also) encode human personality stereotypes." *SEM@NAACL-HLT, 2019, pp. 205–211.
9. F. Durupinar Babur, K. Wang, A. Nenkova and N. Badler. "An environment for transforming game character animations based on nationality and profession personality stereotypes." Proc. AIIDE, published poster, 2016.

One day, around 2006, undergraduate DMD student Meng Yang sat in my office. She was showing me her 3D model of the Byzantine era "Great Mosque of Córdoba" which she had created for her Art History class with Professor Renata Holod. In the middle of her presentation, one of the Engineering School Development Officers, Eleanor Davis, walked past my office unannounced with a Penn alumnus, Harlan Stone. She was escorting Stone around the School as a potential donor. Stone came into my office and excitedly watched Yang's 3D model spin on her laptop screen. Stone had been an Art History major at Penn and loved to see overt connections between the visual arts and engineering. Within a year, Stone donated funds toward our SIG Center renovation. We named the large motion capture space in his honor.

Stone's fortuitous visit spawned a new and broader perspective on my research mission. The HMS Center could branch out beyond purely virtual human work and encompass new projects in digital visualization, especially of cultural artifacts. Capitalizing on Yang's Córdoba model, Joe Kider and I met with Professor Holod to see if we could portray the interior lighting more realistically. Originally, the mosque interior spaces were lit with elaborate hanging fixtures called polycandela, which held numerous oil-fueled glass lamps. I invited a colleague to visit with us for a week: Professor Alan Chalmers from the University of Warwick in the U.K. Kider and I were greatly impressed with Chalmers' work on recreating illumination based on the spectra of natural fuels, such as oil, wood, or wax. Kider took on the project of re-synthesizing oil-based candlelight in the mosque space and signed on two DMD seniors to help, Rebecca Fletcher and Nancy Yu. The SIG Center conference room became a lighting study lab, as Kider set up numerous empirical studies of how variously-shaped oil-filled lamps created illumination. The most surprising result is that although the flame creates upward light, the addition of a layer of water on

N. Badler, *On Raising a Digital Human*, Synthesis Lectures on Computer Science, https://doi.org/10.1007/978-3-031-63945-6_41

top of the oil refracts the flame light downward! Glass oil lamps hung from the ceiling could thus provide quite adequate floor-level lighting [1].

This graphics project got me thinking. Perhaps there could be a broader research component to the SIG Center focused on interesting visualization projects. Working with Ginny, I had already made a foray into cultural reconstruction with the SITE system in 1978. More recently, I had been co-teaching a class with an Anthropology colleague Professor Clark Erickson called *Visualizing the Past, Peopling the Past*. (More on that later.) With the publication of the mosque relighting paper, I began to formulate a post-HMS Center research strategy. We received small but welcome incremental gifts from Harlan Stone and other Penn alumni, but I wanted an umbrella organization to support possible larger projects. Understanding how universities work, I would create a new research Center, housed in the SIG Center space. I called it the Digital Visualization or ViDi Center, as a play on the famous line from Julius Caesar: *veni, vidi, vici* (I came, I saw, I conquered).

The ViDi Center launched in 2014. We held a full day symposium kick-off event and invited leading academic proponents to speak on integrative research in the digital humanities: Ming Lin from the University of North Carolina at Chapel Hill, Holly Rushmeier from Yale University, and Stephen Seitz from the University of Washington. Their talks clearly established that the ViDi concept was indeed well-founded. However, depending on one's point of view, ViDi was a success or a disappointment. On the success side, we built a robust summer undergraduate research program, engaged students in novel and compelling visualization projects, and published enough research papers with undergraduate participants that I received a University Undergraduate Mentorship award in 2017. I am indebted to my Ph.D. student and lab manager, Joe Kider, for firmly establishing that path by the time ViDi started.

Any failure to raise significant research funding for the ViDi Center must be attributed to my lack of suitable proposal writing. It was much less expensive to utilize undergraduates, even when we did pay them a decent summer stipend. Moreover, undergraduates and some Master's students had the right skills and incentives to produce quality models and animations for their own resumes. Because most of my last Ph.D. students worked on the Army's human–robot interaction marketplace project and had graduated by 2015, I did not have to find support for anyone. That alone was a disincentive to write a big grant, but it was also due to some latent laziness that started creeping up on me as 2020 approached.

The ViDi Center led several projects between 2014 and 2022. Our first project created an animated model of the Rittenhouse 1771 Orrery, a mechanical solar system device housed in the Rare Books section of the Penn Van Pelt Library. DMD student Sally Kong led an all-women team of DMD, fine arts, and computer science students: Jun Xia, Isabela Rovira, Rachel Han, Karen Her, and Wenli Zhao. One particular challenge they addressed: they had no direct access to the device, which was kept in a sealed glass case. They had to work from photographs and documentation from a prior restoration.

Our next project was a 3D virtual reconstruction and walkthrough of the ENIAC computer, as it existed in the Moore Building circa 1945. Bill Mauchly, son of one of the ENIAC inventors, invited us to resurrect this particular configuration. After being installed and tested at Penn, ENIAC was disassembled and moved to its home at the Army Aberdeen Proving Grounds. DMD student Isabela Rovira produced the Moore ENIAC models and video for its 70th anniversary celebration. The University of Pennsylvania Archives held useful diagrams and photographs to help reconstruct the in situ configuration.

I've already mentioned the 3D model of the Reading Terminal Market we used for the thermally-influenced crowd simulations. ViDi sponsored the summer students. I tried making some overtures to the Reading owners to see if they might like to use it for marketing, virtual tours, or interactive maps. That inquiry went nowhere. I guess they were happy enough using real photographs and videos of the place. I can't blame them, though. Our 3D model looked barren without the colorful crowds of real people who enjoy it in reality every day. Our human models just couldn't meet that standard.

A little more than a decade after I met Harlan Stone and received his strong encouragement to start the ViDi Center, he would become the principal contributor toward the new Penn Data Science building opening in 2024. Instead of "Stone Hall," he asked that it be named in honor of the Penn President at the time: Amy Gutmann. (It would also be Philadelphia's tallest mass timber construction, which would have made a "Stone" title even stranger.) An interesting coincidence: Stone's initials are HMS, the same as my first research center's name housed in the SIG space. The odds here are 1 in $26^3 = 17,576$; not the lottery, but still rather slim!

My ViDi efforts returned to archaeology in our next project. We addressed an interesting combination of domains, covering computer graphics rendering, natural material reflectance properties, and human factors. With such breadth, and a personal connection as well, it deserves deeper elaboration.

Reference

1. J. Kider, R. Fletcher, N. Yu, R. Holod, A. Chalmers and N. Badler. "Recreating early Islamic glass lamp lighting." Proc. International Symposium on Virtual Reality, Archaeology and Cultural Heritage (VAST), 2009.

Archaeological Lighting

Being married to an archaeologist naturally enabled cross-connections worthy of exploration. Ginny and I first collaborated on the SITE system in 1978, which merged object databases, 3D spatial data, and interactive computer graphics. My first Ph.D. student, Bulent Özgüç, had been instrumental in obtaining Ginny's access to an active excavation in Turkey in the 1990's. In the early 1970s, her Ph.D. supervisor, Professor T. Cuyler Young at the University of Toronto, had excavated a Near Eastern site called Godin Tepe in Iran. Her task, in the most general terms, was the description and analysis of the excavated materials from the period around 3000 B.C.E. She had firsthand access to the artifacts, mostly pottery sherds, stored in Toronto's Royal Ontario Museum. The site of Godin Tepe was a mound, built up as mud brick structures collapsed as successive occupants constructed new buildings and functional spaces on top of older ones. Her study levels were the last and deepest ones excavated, meaning they were the oldest. Fortunately, the scale of the excavation was modest enough that around 2005, some of my students built a full 3D graphics model of the extant walls and rooms. They animated a dizzying eye-level flythrough of all the rooms, which nonetheless gave a viewer a sense of presence, scale, and complexity.

Against this backdrop, Ginny made some exciting discoveries. Based on her archaeological hunches, she found evidence of ancient wine- *and* beer-making residues on uncleaned pottery sherds. Working with a colleague from the University of Pennsylvania Museum, Dr. Patrick McGovern, and a chemist, Dr. Rudolph Michel, they chemically verified that the pottery residues excavated at Godin Tepe were diagnostic evidence for beer and wine [1]. For a short while, they were in the *Guinness Book of World Records*, until further investigations at other regional sites produced older samples. McGovern subsequently focused his career on investigating ancient residues.

Something else about the Godin Tepe artifacts intrigued her. Archaeologists can only examine what remains on a site and, from that, try to infer human behaviors. Although there were hundreds of pottery sherds, there were few so-called *small finds*. These are items such as beads, weights, pins, and bits of metal. All are useful evidence of human occupation, but especially signify technology, ornamentation, and valued goods. She was surprised that so few of these items were extant. Generally, they are robust (such as metal pieces) or precious enough (such as jewelry) to be kept safe or else stolen, as the case may be. Ginny felt that the Godin Tepe site was abandoned quickly and that any precious items were gathered and removed in haste. The few remaining small finds were there because they may have been *overlooked*. Even in broad daylight or by the light of a hearth fire, it is conceivable that some parts of a room were so dark that even small objects in plain sight could have been hard to see.

Ginny knew from the excavation notes where the small finds were found, but on a 2D plan of the site, it was not clear whether they would have been visible under normal lighting conditions. An additional complication is that there was no way of knowing what time of day, what day of the year, or even what exact year the site occupants left. Computing shadows from a light source was one of the earliest features added to computer graphics renderings. However, at Godin Tepe, we had no idea where the sun would have been at the time of abandonment. In preparation for our lighting studies, CGGT Master's student Megan Moore upgraded the existing Maya site and architecture 3D model with completed hearths, walls, doorways, and ceilings.

After Joe Kider graduated from Penn with his Ph.D. in 2012, he held a four-year research position with Don Greenberg at Cornell University. Kider's colleague at Cornell, Dr. Bruce Walter, had built and operated one of the few working instruments to measure a material's Bidirectional Reflectance Distribution Function (BRDF). The BRDF is the empirical description of how a surface reflects light, and includes color, smoothness, texture, and direction-dependent terms. Walter's device could take the measurements needed to establish a material's BRDF. Although we had 3D models of the Godin Tepe architecture, we did not know the reflectance properties of the walls and floors. Without some handle on those, we couldn't do any realistic lighting studies. In a room with a dirt floor and mudbrick walls covered in dirt-colored plaster, most illumination will be diffuse, or scattered, like matte paint. Unlike harsh, sharp-edge shadows, diffuse surfaces bounce light more uniformly and tend to fill in shadowed areas. Thus, we had three issues to resolve: what were the reflectance properties of the mudbrick materials at Godin Tepe, what were the possible light sources in a room, and where might the sun rays directly enter a space over the span of an entire year circa 3000 B.C.E.?

Walter's reflectance instrument gave us the first answer. Ginny borrowed mudbrick samples from the Royal Ontario Museum collection. Walter measured their BRDFs both as dirt and as wetted then dried dirt, or plaster. These BRDFs were then associated with the 3D model floor and wall surfaces, respectively. The second question about internal light sources was easy: the rooms that contained small finds each had an obvious hearth

Fig. 42.1 Godin Tepe room with hearth fire and one specific sun position illumination. Even so, the portion of the floor with the small find (circled) is very dark

that would have been the source of any internal illumination. Moreover, since the hearth burned wood, we could use actual wood combustion spectra provided by Alan Chalmers as the light source. No electric lights back then. We made a reasonable assumption, based on contemporary structures and common sense, that the ceilings were solid.

Kider approached the third question, sunlight, by taking a conservative, global perspective. He simulated all possible sun positions over a 3000 B.C.E. year and integrated the illumination results over an accumulation grid superimposed on the floor. At Godin Tepe, the rooms of interest had narrow doorways, thick mudbrick walls, and probably a timber roof. Sunlight could only enter through the doorway or a window, if the room had one. Even over the course of a full year, sunlight barely penetrated much beyond the door threshold. Most illumination arose from the internal hearth fire.

We were able to restrict our attention to two rooms with small finds. Using a proven rendering system called Radiance [2], Kider generated images of the small find rooms with maximum hearth illumination and the full-year sunlight floor maps. Radiance output provided a human perceptual measure of visibility in *lux* units. A typical human reading illumination range is 300–500 lux. Ginny's small finds were located in the darkest parts of these rooms, with lux values between 4 and 22. Indeed, even in strong daylight and a blazing fire in the hearth, these small finds would likely pass unnoticed. In one of the rooms, the small find was adjacent to the doorway, in a dark corner with only 9 lux illumination (Fig. 42.1). In sunshine, the visual contrast between the bright doorway and the dark corner would have been even more pronounced. Ginny's observation appeared to be true [3]. We could only make this assertion through an almost accidental alignment of skills, people, and equipment between Penn and Cornell. Our illumination study exemplified interdisciplinary ViDi research.

We thought about trying to populate the Godin Tepe site with animated people, but we just didn't know enough about what went on there to try anything definitive. It had the appearance of a fort, a clearly defensible entrance, a distribution room for slingballs, and a possible winemaking area [4]. However, I already knew about Anton Bogdanovych's virtual recreation of daily life around 3000 B.C.E. in Mesopotamia [5]. There was no reason to compete in the same time and place. I was already redirecting my energy to re-animate another part of the ancient world, South America.

References

1. R. Michel, P. McGovern and V. Badler. "The first wine & beer." Anal. Chem. 65 (8), 1993, pp. 408A–413A.
2. G. Ward. "The RADIANCE lighting simulation and rendering system." In Proc. Annual Conference on Computer Graphics & Interactive Techniques, ACM, New York, NY, 1994, pp. 459–472.
3. V. Badler, J. Kider Jr., M. Moore, B. Walter and N. Badler. "Accurate soil and mudbrick BRDF models for archaeological illumination rendering with application to small finds." Eurographics Workshop on Graphics and Cultural Heritage, Graz, Austria, 2017.
4. V. Badler. "A Chronology of Uruk Artifacts from Godin Tepe in Central Western Iran and Implications for the Interrelationships between the Local and Foreign Cultures." In *Artefacts of Complexity: Tracking the Uruk in the Near East*, J. N. Postgate (Ed.), Iraq Archaeological Reports 5, British School of Archaeology in Iraq, 2002, pp. 79–109.
5. A. Bogdanovych, J. Rodriguez, S. Simoff and A. Cohen. "Authentic interactive re-enactment of cultural heritage with 3D virtual worlds and artificial intelligence," International Journal of Applied Artificial Intelligence, Special Issue on Intelligent Virtual Agents, Vol. 24(6), 2010. pp. 617–647.

Maybe I'm slightly claustrophobic, but for as long as I can remember, I liked to keep my office door open. This can lead to interesting situations. For example, I once encountered a "walkaway" from the mental facility at the nearby Penn hospital. He came to my open office door and demanded that I change the grade I gave him in some class. Well, it wasn't any of my classes. He seemed to claim that someone from neighboring Drexel University was the culprit who gave him a bad grade. I said I'd call up the professor and see what I could do. I called the Penn police instead and they gently escorted him away.

That was an anomaly, of course. One day, a gentleman came to my door asking for me by name. I said yes, I was Norm Badler. He said that someone in the University Museum's Information Technology office had sent him over to see me because I was the computer graphics person. My visitor was Clark Erickson, an Anthropology Professor and Curator of the South American Section of the University Museum. The Museum is literally one block away (and next door to the Hospital). He wanted to talk to me about applying computer graphics to a project he had in mind: drawing the pottery cross-section illustrations that were standard in archaeology and anthropology publications. These illustrations required skilled drafting, making them potentially tedious, time-consuming, and thus expensive tasks.

Erickson thought that perhaps I could help him figure out how to automate this process, or at least do it with some technological assistance. I gently interrupted him and told him he'd come to the right place. It so happened that because of Ginny's Godin Tepe archaeology, I had already helped her draw hundreds of pottery sherds. The process began with a sherd that had a bit of object rim remaining on it. By placing it upside-down on a plastic mat imprinted with a series of concentric circles, one could measure the radius of the entire pot from the sherd and obtain a good estimate of what size and kind of

N. Badler, *On Raising a Digital Human*, Synthesis Lectures on Computer Science,
https://doi.org/10.1007/978-3-031-63945-6_43

pot it might have been. This process, called stancing, was something I had been helping Ginny with for a long time. She would hand draw the sherd profile and then scan in the drawing. Using a software application called CoralDraw, she digitized the outline by carefully tracing the outline point-by-point. What she didn't know is that CoralDraw used a graphics technique called *b-splines* to draw curves. It was tedious to do closely spaced points along the outline. Instead, one needed to take advantage of the application's ability to create smooth curves from only a few salient outline points. I understood how it worked, so I offered to take over these drawings from her. She agreed and thereafter that became my job. I got quite good at it. I traced all the excavation plans and sections as well. All the drawings that Ginny needed for her publications began as my CoralDraw files.

Erickson's idea involved using computer graphics software to draw sherds even more automatically. I told him about our mostly unsuccessful attempts to automate the stancing process using the Polhemus motion capture system in the 1980s. We couldn't both touch the sherd with the wand *and* keep it absolutely stationary. As we talked for a couple of hours, it became clear that Erickson's real interest was anthropological—how a piece of pottery was used—rather than just knowing its type or shape. As a young archaeologist, he had certainly drawn many pots from sherds. The perspective he brought from anthropology was a humanistic one. He felt that just drawing pottery, or showing them in illustrations without their human context, was leaving out the most important part, namely, their relationship to how people lived. This perspective was completely in line with Ginny's own view. In fact, this was the primary reason she switched from Art History to Archaeology for her graduate work. It wasn't the pottery itself, say as a museum object, which interested her. What was important was the found context (such as the small finds in the dark corners!), how it was used, and especially how it was made. She even learned to throw pottery on a wheel to gain a personal understanding of the ceramic creative process.

By the time our meeting ended, I was receptive to Clark's idea of creating a course at Penn that combined our individual interests in computer graphics and anthropology. We wanted to elaborate on the process of visualizing the past, whether it be in artifacts or architecture. We also wanted to address the human element: showing the people who created, lived with, and used these artifacts. These ideas motivated the course that we taught many times over twenty years which Erickson named *Visualizing the Past; Peopling the Past*. We started to offer it every other year, and ultimately, we ended up teaching it once a year to an interested clientele of students across the university. Our students were an eclectic lot: not just engineers and not just anthropology students. While I had co-taught courses in the past with other computer scientists, notably my natural language colleague Bonnie Webber and my graphics colleague Stephen Lane, this was the first time I co-taught with someone totally outside of my expertise. I learned along with our students. Likewise, Erickson absorbed a firsthand view of computer graphics technology, including 3D modeling and animation.

Erickson had already worked closely with an illustrator to help interpret and recreate scenes from fieldwork in South America. Erickson also had extensive curated materials on other artists who illustrated habitats and lifestyles throughout human history. He was especially interested in the imaginative, constructive, creative processes and the implicit biases that artists brought into that enterprise, working from minimal source materials. One of the early assignments in our course was to have each student study a particular artist's portfolio of illustrations, usually of some period, site, or culture. Each student presented and commented on the artist's works with an emphasis on what was based on reality and what was invented.

All our artist exemplars were "classical" creatives working in oil, watercolor, or even cartoon media. We wanted our students to view this through a computer graphics lens. We added this material in two ways. First, I had to teach computer graphics fundamentals without having them understand computer programming. After some unsuccessful attempts to integrate more computer science into the course, we established a very workable model where students learned to construct 3D models in Autodesk Maya in lab sessions once a week during the first third of the course. We soon found an excellent lab leader in DMD student Josh Nadel, who had taken our course previously. Nadel and I developed a series of assignments that students completed in class. The idea was to get them to use Maya successfully in a supportive atmosphere. DMD student TAs circulated among the class during the lab, helping students navigate the sometimes esoteric world of 3D modeling. These assignments were required but not graded. All students acquired adequate modeling skills. In fact, they ultimately appreciated it as a tangible result of taking the course: not only were they examining artifacts from the past, but they were actually learning a generalized modern and marketable skill of 3D modeling as part of the course.

One of the fortuitous course constraints was that we had no readily available 3D scanning device. Students could not just place an object on a turntable and push a button to obtain a 3D model. Our students had to think about an object's construction so that they could model it. The University Museum archives provided a variety of authentic artifacts that our students could examine, photograph, and model from scratch using the Maya tools. The students started by modeling a pottery jar as a surface of revolution directly derived from a photographic profile view. Then, we showed them how to apply decorative textures, handles, and lugs to the 3D model. After they completed the pot, we had them model a generic dugout canoe. This had its own set of interesting problems, such as making cut-outs, handling asymmetries, connecting benches, and applying appropriately oriented wood textures. With these two modeling projects behind them, they were ready to tackle any kind of artifact that they might find interesting for their final project.

Indeed, this introduction enabled numerous final project models that included fishing equipment, bows and arrows, religious icons, culture-specific items, baskets, clothing, rafts, and architectural features. One of the most rewarding aspects of this course was the empowerment we gave students to look at an unfamiliar object and to think about how it was made. That engaged them in a humanistic view of the past, which I first learned from Ginny. Somebody actually put time and effort into creating the original object with a particular function or role in mind. Our students did the same, but on a computer. This was the anthropological lesson behind learning 3D modeling.

Erickson brought an extensive final project requirement to the course. I came to appreciate his methodology. He insisted not only on a *final* report, but also that it also methodically document all the steps, including any false starts or dead ends, that they may have taken along the way. As they were building, say, an object in Maya, he wanted them to save screenshots of the work in progress and add a written description of its import. Their final reports, therefore, became documentation of their own learning process. Moreover, Erickson encouraged correction and revision in the written document as many times as needed. The idea is that the student revises to the grade they want, not just the grade initially assigned. Writing quality improves, and the student has tangible incentives and benefits from the feedback. Erickson was experienced in descriptive expository writing, having taught introductory anthropology for many decades. I learned from him, too.

We had a fair share of DMD students taking our class over the years. It was an especially good choice for freshmen, so we could emphasize interdisciplinarity between anthropology, art, and computer graphics right from the start. We had students from across the University, too, and we welcomed them into our world. They did not have to have any prior computer skills. Few, except for some of the DMD students, had even done any sort of 3D modeling. Programming was not required or needed. My favorite example of the course's success was an English major, Lynn Oseguera, with obvious computer phobia. She hardly even wanted to open her laptop. As was expected, she did the Maya modeling assignments in class. If needed, she had readily available assistance. For her final project, she chose to model body tattoos common to indigenous Amazon people. She also wanted to model the process of applying the tattoos. An inked roller created a repeating pattern as it rolled along the arm (Fig. 43.1). She modeled these patterned rollers, too. Even though we knew she had some Maya help from the DMD TAs, her final project was amazing. She did what she set out to do: she animated applying face and arm tattoos and animated the human models to show off her handiwork. I don't know who was happier with the result, her or us, but we think she gained much-deserved confidence along the way.

Erickson's expertise in South American anthropology informed our topic choices for the class. Over the years, we gravitated toward two principal pre-Columbian communities: the Amazon Baure culture in Bolivia and the Inca civilization in Peru. Both took advantage of their respective landscapes to flourish. The Baure had to contend with and harness the Amazon basin's annual flood-to-drought cycle to provide a year-round food supply of plants and fish. They lived in villages on raised forest islands and created an extensive network of canals and land bridges for transportation. They cultivated a variety of foodstuffs and developed fish traps and weirs to hold receding flood waters through the dry season. Baures artifacts include tools, bows and arrows, a wide variety of baskets and fish traps, hammocks, and dugout canoes.

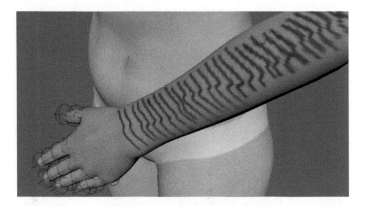

Fig. 43.1 Rolled-on Baures arm tattoo by Lynn Oseguera

Several students from our class and some additional interns launched an ambitious summer project to create a unified animated reconstruction of the Amazon Baure culture. DMD students Myles El-Yafai and Josh Nadel led the team. Over the summer of 2017, our students produced a two-minute animation showing the Amazon flooding, the fish weirs and capture baskets, a forest village of procedurally-generated thatched houses, and a variety of pots and baskets. They also motion captured themselves performing appropriate behaviors, such as rowing a canoe, fishing, or telling stories around a fire (Fig. 43.2). Properly attired human models were adapted to resemble Baures people illustrated in contemporary Spanish accounts of the time. Vegetation appropriate to the Amazon jungle created the right physical context. Animal and insect sounds Erickson himself recorded in the Amazon became the soundtrack. The students elected to render the video using the Unreal 4 game engine, so the graphics were beautiful [1]. By the end of the summer, we had a video everyone could be proud of [2].

Erickson showed the Baures video at a conference of South American anthropologists. The audience insisted that he show it a second time, at least. I also showed it at the Symposium on Virtual and Augmented Reality in Brazil during my Keynote talk in 2018. No one had seen anything like it before. Our interdisciplinary course now had a highly visible "product." Both Erickson and I were invited to speak at a special "Penn Academy" event in Beverly Hills, California, in April 2019. Held every two years in various venues, this was a show open only to top Penn donors. Erickson and I were on the highly selective program introduced by Penn President Amy Gutmann. Our talks were among other significant speakers offering a cure for blindness or specialized working dog training. Our Baures video clearly conveyed our interdisciplinary messaging. We felt honored that our collaboration over the years had educational and visual consequences.

Unfortunately, sometime between the middle of August 2017 and the start of classes later that month, someone erased the hard drive holding all the 3D models, scripts, and Unreal engine assets. No one had any idea who did it. No one had made a back-up. The

Fig. 43.2 Baures models and a basket

files were simply gone. All we have now is the completed video. Al-Yafei and Nadel took it especially hard, but there was nothing to be done. They both had to move on.

After the Baures studies, we switched to an almost polar opposite culture, the Inca. We had highlighted them in earlier versions of the course, but with the Baures visualization successes, a revisit was in order. The best surviving remnant from that earlier version was Adam Mally's Maya model and animated flythrough of the Sun Temple. That complex dominated the landscape around Pachacamac, an extensive Inca ceremonial center situated on the coast in Lima, Peru. In summer 2019, we began to add models to the site with the intention of building out as much of the extensive architecture as possible through classwork. We had a decent start and were determined to attack the entire site in earnest as an undergraduate summer project in 2020.

Our major goal in 2019, besides the architectural models, would be to populate the site. Aline Normoyle, who had completed her Ph.D. with me in 2015, was already my research specialist on an unrelated NSF project. She had experience in building agent activity models with temporal, capacity, and flow constraints [3]. Although I had no role in her effort, I liked this work very much. I felt it could apply to evaluating use and access hypotheses in ancient sites such as Pachacamac. Like Allbeck's CAROSA model, Normoyle's agents occupied a building represented as a graph of interconnected nodes. However, Normoyle's agents behaved stochastically rather than individually; that is, they responded in a probabilistic fashion to external constraints rather than following an internal schedule. Thus, the environment, such as a mall with a multiplicity of stores and restaurants, would demonstrate global satisfaction of timing and capacity constraints rather than hoping that individuals would independently make good decisions.

We believed that this approach might work to characterize possible human flows among and within the Pachacamac site. Moreover, we expected that the site itself was engineered to steer crowds into narrow passageways, perhaps to restrict certain groups of people or create more orderly access. From historical and contemporary sources, we believed that processions would also be an important crowd feature. I began to collect processional videos from YouTube and documented a new crowd model tailored to processions. One of my DMD students, Kristen Chow, began work on the model. We made some headway but did not have time to get the implementation to a sufficient level of usability [4]. When Chow graduated, she received a job offer from Walt Disney Animation Studios to work as a crowd specialist. Her email to me announcing this was just so wonderful and honest. It ended with "I am both thrilled and very frightened." She did accept the job and made major contributions on her first two Disney films, *Raya and the Last Dragon* and *Encanto*. Fear vanquished!

As we were ramping up to tackle the Pachacamac site simulation, the COVID-19 pandemic struck in early Spring 2020. Suddenly, we could no longer hold any in-person summer work in the SIG Center. Rather than give up and renege on our summer research offers, we decided to proceed and meet remotely as needed via Zoom. I asked Normoyle to continue to manage the project. She was very good at supervising the four undergraduate interns: DMD students Felicity Yick and Samantha Lee, Penn computer science student Emilia Soto, and CalTech summer researcher Shir Goldfinger. The team spread over three continents but managed to find a common time to have Zoom meetings. They built out much of the known architecture of Pachacamac. We had an excellent team, but separation, graduations, and other obligations slowed progress considerably. Normoyle kept it going for a while, but even she had other obligations to fulfill as a new Assistant Professor at Bryn Mawr College. After teaching *Visualizing the Past; Peopling the Past* once more (remotely) in Fall 2020, both Erickson and I retired from Penn in July 2021. It's easy to blame the pandemic for any further lack of progress, but in reality, I think we did what we could. The remaining effort would have been substantial and of a sort that might have only a purely hypothetical benefit. We weren't prepared to extend our commitment that far into the future.

References

1. C. Erickson, E. Al Yafei, J. Nadel, Y. Victor, I. Ogiriki and N. Badler. "Recreating Pre-Columbian life in the Baures region of the Bolivian Amazon." Proc. 20th Symposium on Virtual and Augmented Reality (SVR), Foz do Iguaçu, Brazil, 2018.
2. Baures animation video. http://cg.cis.upenn.edu/VTP/recreatingbaures.html. Accessed November 9, 2023.

3. A. Normoyle, M. Likhachev and A. Safonova. "Stochastic activity authoring with direct user control." ACM SIGGRAPH Symposium on Interactive 3D Graphics and Games, 2014.
4. K. Chow, J. Nicewinter, A. Normoyle, C. Erickson and N. Badler. "Crowd and procession hypothesis testing for large-scale archaeological sites". Workshop on Modeling and Animating Realistic Crowds and Humans (MARCH), 2nd IEEE International Conference on Artificial Intelligence and Virtual Reality (AIVR), 2019.

Having collaborated on eye movement projects with my older son, Jeremy, it is only fair that I include what my younger son, David, precipitated. Although both attended Penn, Jeremy followed a bioengineering major that eventually led to his Ph.D. in neuroscience. David took a different path. In 1998, he entered Penn's prestigious Management and Technology undergraduate program, where students receive degrees from both the Wharton School of Business and the Engineering School. David opted for a computer science major.

Tradition in the Engineering School allowed faculty to personally present family members with their diploma at Commencement (graduation). I always felt quite awkward at graduation ceremonies; perhaps it was fear of boredom or of falling asleep on stage. I assiduously avoided all of them since high school, until I wanted to present Jeremy with his Penn Engineering degree in 1995. I did attend that one. In 2001, as Associate Dean, I could not escape attending Commencement every year and standing in the graduate reception line on stage. I even had to give a short speech at the event. My anxieties were growing.

At an academic meeting that I had to attend in a local hotel, I took a much-needed bathroom break. While standing to relieve myself at a urinal, I relaxed, and the idea of a lifetime popped into my head. We had been building interactive graphics systems in my lab for decades. What if we made graduation ceremonies *visually interactive*, too?! We motion-captured people for animations, but for a ceremony, we only needed to capture their *identity*. We would give each graduate a card with a unique barcode and then scan the card as their place in line came up. The coded identity would trigger the display of a pre-stored webpage with the graduate's name, picture, major, and a short message.

N. Badler, *On Raising a Digital Human*, Synthesis Lectures on Computer Science,
https://doi.org/10.1007/978-3-031-63945-6_44

Excited about this new way to run an otherwise linear ceremony, I coined the name *MarchingOrder*. I pitched the idea to David. He thought well enough of the general idea to contact another Management and Technology student, Tyler Mullins, to collaborate on realizing the technology for their required computer science senior project. I wrote up MarchingOrder on a Penn Technology Transfer Disclosure Form. This was Penn's way of securing intellectual property rights in case an idea reached commercialization, patent, or licensing stages.

David and Tyler ordered an inexpensive barcode scanner and began to design and code the data structures and interactions. A DMD student, Matt Uffalussy, helped with the graphical display layout. The Engineering School agreed to let them test and run it live at the May 2001 graduation ceremony. They already had a Jumbotron TV display for the event, which projected a live video camera feed so that the audience could at least see what was going on. Now, however, everyone could also see synchronized and individualized MarchingOrder webpages. The display nominally contained a photo of each graduate, their readable name, and the graduate's own humorous or appreciative message. It was dynamic. The audience now had something to do, at least until their own source of pride was called up.

In 2002, Penn Engineering ran it at David's and Tyler's own Commencement. In addition, MarchingOrder entertained at several other Penn Schools (each had their own ceremony), and branched out externally to the University of New Mexico. Although David soon turned his attention elsewhere, Tyler took over MarchingOrder and made it his life's work. It is now much more feature-rich, making it the undisputed leader in the graduation event space. Our *Jack* software undoubtedly touches many individuals, such as workplace designers, evaluators, or satisfied occupants. However, as of 2023, MarchingOrder had already served over 1500 ceremonies covering 2.5 million graduates and orders of magnitude larger audiences in attendance. This success makes me happy: it is a tangible legacy of my administrative role. And, yes, I was able to hand David his diploma personally. In 2002, the graduates themselves were not yet virtual!

Publicity

For many professors, recognition within one's peer group is pretty close to the meaning of life. We do research, write papers, give talks, and hopefully glean accolades (or at least recognition) from colleagues who cite our papers. I had been on the Penn faculty a few years before I realized that my actual role was selling myself and hawking my wares of ideas and graduating students in as many venues as possible. This realization changed how I approached my faculty position: I was a salesman who sought to raise money to continue and grow my enterprise. I became quite adept at obtaining research funds to support the students and equipment in my lab. I was even more fortunate to have a marketable product in the *Jack* software. I started calling myself "Salesman *Jack*" as I traveled with increasing frequency to spread interactive digital humans into government sites and human occupant, workplace, and vehicle design industries. The irony of this self-induced moniker was that I had been determined *not* to follow in the footsteps of my salesman father. I even scored lowest in the "salesman" category on a vocational profile questionnaire in high school. Fate had other plans for me, it seems.

To sell, one needs a marketing strategy and publicity to advertise one's products. My strategy was to be as visible as possible, not only in classic academic conference venues but also in smaller, industry-focused communities. Being attentive to industry needs and bringing along viable technological solutions opened many industrial doors to *Jack* and the associated research projects that it funded. Industry itself has marketing needs, so we soon found that they were helping promote us as well. In addition to the usual academic conference fare, we started appearing more often in popular media.

Bubbleman and Bubblewoman first captured the attention of major magazines:

- *Popular Mechanics*, "This dance is mine". Bubbleman image. April 1979. p. 21.

N. Badler, *On Raising a Digital Human*, Synthesis Lectures on Computer Science, https://doi.org/10.1007/978-3-031-63945-6_45

- *American Way* (Airline Magazine), "Computers have Designs". July 1980. pp. 59–62. Bubbleman image, p. 62.
- *Science*, "Video Graphics and Grand Jetés". May 1982. pp. 24–33 (Bubbleman, p. 31).
- *Self*, "The first 3-D computer exercises". September 1984. pp. 186–188.
- *Millimeter*, "Figuring it out". April 1988. pp. 47–54. Included Bubblewoman in an otherwise unpublished bathtub scene image (p. 52).

By 1987, there was a healthy Hollywood-based virtual human scene. We were included in this community:

- *VDI Nachrichten Magazin* (Germany), "Die digitale Schöpfung". October 1987. pp. 48–55. Includes Bubblewoman and Tempus figures adjacent to the famous Robert Abel Superbowl ad for cans, "Sexy Robot."

Jack opened up new publication venues since our customers found it beneficial to tout their own application successes. CAD was an already well-established technology. *Jack* was promoted as a crucial new component for human-centered product design digital mock-ups. Some of the major venues were:

- *Scientific American*, "Human Spec Sheet". November 1991. pp. 132–133. This illustrated *Jack's* visual view capability, work initially done for NASA Ames Research Center.
- *Forbes*, "But can she act?". December 10, 1990. pp. 274–278. This mentions *Jack* at Deere in the context of Hollywood-style virtual humans. This was a good cross-section of the state-of-the-art of that segment at that time.
- *NASA Tech Briefs*. September 1992. p. 148. This contained several images and a good summary of the *Jack* system and its interface flexibility.
- *Focus Magazine* (U.K.), "Jack and Jill: citizens of virtual reality" August 1994. Includes several images; apparently promoted by GMS, our U.K. *Jack* distributor.
- *Newsweek*, "Software au Naturel". May 8, 1995. pp. 70–71. Shows Deere use of *Jack* in 3 background images.
- *Mechanical Engineering*, Cover image: "Touched by Almost Human Hands" and "Role Models". July 1999. pp. 44–49. (Engineering Animation, Inc.) Includes other ergonomic human model system images for comparison.
- *Simulation*, Cover image. September 1999. p. 143 (Engineering Animation, Inc. image)
- *New York Times*, "Jack is put through the wringer, so you won't be". May 11, 2000. pp. G1, G8 includes photo of me on Ben Franklin bench sculpture on Penn campus; John Deere applications.
- *Fortune*, "Hot Technologies; Virtual People that Help Design Products". June 24, 2002. pp. 162(H-N). Various *Jack* and *Jill* applications with images.

Our NSF face workshop spawned an article in the *New York Times*: "Japanese Put a Human Face on Computers," June 28, 1994, C1, C11. Unfortunately, this precipitated a hate letter from someone who thought computers were going to replace people.

Our more AI-ish work on motion planning, and especially Cassell's project animating conversation with Gilbert and George, received good mentions:

- *Discover*, "First Stab at a Virtual Human Being", July 1994. Gilbert and George dialog p. 4 and article "Virtual Jack" pp. 66–74. Many images of AI work with Bonnie Webber, including *Jack* doing the limbo and the planner-based Soda*Jack*.
- *Harper's Magazine*, "Withdrawing from Reality". August 1994. p. 16. Shows the Gilbert and George dialogue.

Medisim had at least one mention, in a front page article in a newspaper *The Orange County Register*: "Meet Jack, Virtual Man", June 19, 1996. pp. 1, 17. In addition to the Medisim scene, it showed *Jack* in a Deere tractor.

Sometime during the 1990s, I was interviewed by CNN in New York City. It aired, and I thought nothing further of it and received no feedback. A few months later, I was on an international flight returning from Germany. Back then, the airlines had drop-down TVs mounted over the aisles rather than today's individual seat-back versions. I usually don't watch anyway, but a familiar image caught my attention. It was my CNN interview, but it was dubbed in German! I slumped down in my seat, unsure if I would be recognized by the other passengers. Apparently, I wasn't, or else they were too polite to say anything.

Another lengthy interview orchestrated by Philadelphia documentarian Joe Glantz, is online [1]. Glantz included a number of illustrations on the published site. His intention had been to produce a sequel to his first volume, *Philadelphia Originals* [2], but I do not believe a print version materialized.

In 1999, Frank Foster produced a seminal movie with interviews of many early computer graphics pioneers: *The Story of Computer Graphics* [3]. Filming took place at a SIGGRAPH conference and included professional make-up and lighting. Asked to participate, I readily agreed. I knew I would be asked about virtual human animation, and I ad-libbed responses appropriate for the intended public audience. I described the three main computational approaches to animation: manual scripting, physics procedures, and motion capture. Although I do appear in person, most of what I said ended up distilled into the movie's narration delivered by well-known Star Trek actor Leonard Nimoy. He certainly gave my words greater impact with his usual gravitas. And there's a good ego-stroke on the IMDB site where the fortuitous alphabetical listing of "Stars" places me in the same line as Nimoy!

As a bizarre codicil on raising virtual humans, I'll relate a true story about *Jack*, popular culture, and me. Most people don't expect to find themselves mentioned in a novel. I certainly didn't, but that's what happened to me. I never would have known except for an extraordinary coincidence.

Three faculty co-founded the DMD undergraduate program at Penn in 1998: Julie Saecker-Schneider in the Graduate School of Fine Arts, Paul Messaris of the Annenberg School, and myself in CIS. In its early days, Julie, Paul, and I met monthly to manage the program. At one meeting Julie walked in and asked me if I knew that I was mentioned in a mystery novel by Daniel Hecht called *Skull Session* [4]? No, I hadn't known about that! Was it really about me or just a coincidence? No, it was about me and featured *Jack*. No mistake here. How did she find out? Julie was taking a long plane flight to Los Angeles and had picked up a novel at the Philadelphia airport to read on the way. Apparently, she is fond of mysteries, and this novel was recommended in *People Magazine*. Hecht, the author, never contacted me directly; everything he used was gleaned from my webpages. I had a second hit in 2006, when I found out that the mother of one of my former DMD students was fond of listening to books on tape while driving. She happened to listen to *Skull Session* and asked her son if the discussion about "Virtual *Jack*" and "Dr. Badler" was real or not. Although that discussion is fiction, both of us are real.

References

1. J. Glantz. Interview with Norm Badler. http://www.joeglantz.com/Interview_NormBadler.html, 2018. Accessed March 31, 2024.
2. J. Glantz. *Philadelphia Originals*. Schiffer Publishing Limited, 2009.
3. F. Foster. *The Story of Computer Graphics*. https://www.imdb.com/title/tt0210309/. Accessed March 31, 2024.
4. D. Hecht. *Skull Session*. Signet, 1998.

When I arrived at Penn in 1974, the CIS Master's program attracted students from several industry sources in the Philadelphia region. These included established employee educational benefit programs with the likes of Lockheed-Martin, DuPont, General Electric, Boeing, and Univac. There were local government sites such as the nearby Navy Yard, Naval Air Base, and other Army sites in New Jersey. A strong emergent computational emphasis in the various Bell Laboratory research centers in North Jersey also contributed Master's candidates. The old TV classrooms gave Penn a long reach into the technology community. Master's students were generally well educated but felt ready to update their skills with newer computer science courses in computer graphics, computer vision, artificial intelligence, modern programming languages, and databases.

In addition to their coursework, all Master's students were required to write and defend a Master's thesis. For a new faculty member, me, in the hot new field of computer graphics, this requirement was a boon. I did useful and significant research with Master's students, and they had to write up their work, too. Although CIS decided that Master's theses were optional after 1980, they continued to be a vehicle for excellent students who wanted more than just a coursework experience. I supervised thirteen Master's theses from 1980 to 1990, two in the 1990s, two in the 2000s, and then just five more after that. In the post-1980s, those theses covered a wide range of topics, including graphics rendering, user interaction, collision detection, fluid modeling, and human motion models.

Given this sparsity of demand, any student who expressed interest in writing a Master's thesis with me came under scrutiny. One day in 2007, I met with a Master's student, Patrick Cozzi. He worked at a nearby computer applications company, AGI, doing computer graphics. His particular interest lay in graphics rendering of the earth, as one of AGI's core technologies was tracking and visualizing satellite trajectories. Cozzi had some ideas for dealing with the massive amounts of data needed to portray the entire earth, especially data culling algorithms to reduce the amount of surface data processed for each frame of an interactive, earth-scale, 3D map. Although this sounded far from my interests in human motion, computer graphics had long been concerned with data reduction methods to speed up data transfer and rendering. My student John Granieri even did his 2000 Ph.D. thesis on human motion level of detail based on rendered screen size and frame rate.

Cozzi and I agreed on his thesis topic. A year later, he finished, submitted, and graduated. When he handed me the final document, I told him, "Congratulations! You earned a Master's degree the hard way." What I did not appreciate then is that he wasn't done with the topic. Two years later, he asked me if I would read his book draft prior to publication. I said I would be happy to look at it. With his colleague Kevin Ring, Cozzi co-authored a novel treatise on how to display earth-scale graphics [1].

For several years after the book appeared, Cozzi joined the CIS department as a part-time Adjunct Professor so he could teach our graduate-level graphics programming unit (GPU) course. In the early 2010s, GPUs were an uncommon programming target unless one wanted to do computer graphics. GPUs soon became a vehicle for massively parallel programming, and other fields needed to learn about their architecture and programming techniques. Students from computer vision, AI, and natural language processing were flocking to his course and had to learn graphics, literally on the fly, by programming ray-tracers and rasterizers. Quite soon, Cozzi's course earned a reputation as the hardest course in the CIS department, and I had to ask him to dial it back a bit so the non-graphics folks could survive.

Cozzi soon turned the GPU course over to another instructor, but thankfully not because I had chided him about its difficulty. In 2019, his team received AGI's corporate blessing to spin out the earth visualization toolset as its own company, Cesium GS (GeoSpatial). Cozzi became the CEO. Cesium is a leader in open source geospatial visualization. It runs in a browser, so it requires no separate app download. The Cesium ecosystem now includes major technology companies such as Google, Nvidia, ESRI, Epic Games, and Unity.

When I retired from Penn in 2021, I cautiously asked Cozzi if there might be a role for me to play at Cesium. I was quite excited when he welcomed me in. He generously gave me the title *Head of Research*. It was the era of the emerging *Metaverse*—or so Facebook would have us think—but VR was really not a new idea at all. I looked into potentially novel corners where Cesium research might prosper and developed a roadmap with several new directions. In 2022, Cesium hired a summer intern, Kenneth Chen, to work with me on some of these ideas. Chen had graduated from UC Berkeley computer science where he'd been a graphics TA. In 2022, he was completing a Master's degree at New York University.

We settled on one project that matched Chen's significant background and skills. The topic arose from considerations of mounting large-scale performances in a VR setting, such as a concert hall, arena, or stadium. All of these venues could appear in the detailed geospatial models accessed and delivered by Cesium. These venues should contain virtual audiences, but as bare 3D models, none did. If one wanted to watch a live performance through Cesium, a huge amount of streaming data would be required to portray even modest-sized, animated, virtual human audiences. In a real-time setting, data streaming bandwidth severely limits the number of active viewers who could have an avatar present in the performance space. In practice, this means that a live concert, for example, could only entertain at most about 50 avatar attendees. To accommodate more, the live audience must be split into independent groups, each having their own stream. This process is called *sharding*. My idea involved amplifying the audience through *virtual* crowds. There could still be avatars, but the bulk of the background audience would be synthesized in 2D. The code to do this would execute on the client side and not stream as part of the animation. We thought we could build a pilot version. Chen began in earnest.

Chen first collected videos of crowd scenes across a number of venues, including several kinds of sports, rock concerts, pop concerts, and political rallies. He would use machine learning to "understand" the appearance characteristic of each category and then use 2D video prediction and image generation to animate the crowd extrapolated from a single frame example. The surprising result was that the crowd types differed significantly across categories; it was usually possible to ascertain the venue from crowd appearance alone. We concluded that any crowd generation process needed to be aware of the performance category. The second part of the work was prematurely terminated by the end of Chen's Cesium internship. He did use machine-learned video prediction to produce a 10-frame future projection of the audience's appearance. He wrote up both results. The paper was accepted to a conference, but we had to withdraw it when we couldn't get image permissions in time for publication. Chen turned the paper into a poster, so the work is not entirely lost [2]. With the amazing advances in video generation using AI within just the past year, we feel that the results would be greatly improved with better tools.

I have other projects in my roadmap, but what directions I pursue depend on Cesium's own corporate goals. This volume exists only because of Cozzi's encouragement. Originally imagined as a weekly blog post, it grew well beyond a series of self-contained essays into a lifelong exposition. I will be eternally grateful to Cozzi for making this happen.

References

1. P. Cozzi and K. Ring. *3D Engine Design for Virtual Globes*. A K Peters/CRC Press, 2011.
2. K. Chen and N. Badler. "Towards learning and generating audience motion from video." Proceedings of the ACM SIGGRAPH/Eurographics Symposium on Computer Animation, August 2023, Article No. 4, pp 1–2.

Philosophies

Over my 50-year career in academe, I managed to acquire a number of skills that weren't taught in school. In addition to being fortunate to start early in the evolution of computer graphics, I gained an understanding that effective research required management, human resources, and marketing skills. I feel I learned more from my students than they did from me. Especially at a university such as Penn, where Ph.D. students are primarily supported by research funds, being able to obtain and maintain funding is a critical requirement for any multi-year, multi-collaborator enterprise.

I never gave much thought to what the degree "Doctor of Philosophy" meant, other than its obvious roles as educational target, guarded gate, and entrance ticket to academe. The rules of passage vary a bit from place to place and from advisor to advisor, but assessments are in the hands of a committee who set and enforce, to some degree, an expected contribution threshold. Typically, an external thesis reader is engaged for subject matter expertise. This often has the implicit side effect of directly introducing the Ph.D. candidate into a recognized community, perhaps even before a major publication issues. Obtaining a Ph.D. in a technological field such as computer graphics, however, does not mean that one has necessarily taken any *philosophical* point of view. Technical and methodological, and today, even ethical, discussions are appropriate. Defense of one's achievements is required. The successful candidate receives their "golden ticket." The craft of mounting a defensible argument and validating it through implementation, testing, or proof is often as close as we get to a philosophy.

I came to realize that there's another way to look at the Ph.D. It is a purely personal achievement regardless of whether the degree provides an overt admission ticket into the ivory tower. I think this is the main point, which I was better able to articulate after many decades of supervising students: "People get Ph.Ds. who want them." A Ph.D.

© The Author(s), under exclusive license to Springer Nature Switzerland AG 2025
N. Badler, *On Raising a Digital Human*, Synthesis Lectures on Computer Science,
https://doi.org/10.1007/978-3-031-63945-6_47

must be perceived and internalized as a goal for its own sake. I learned that my role as an advisor was to help the student find their path to the degree. If they wanted it, we found a suitable problem to study. Sometimes these came from me and my research directions and obligations. Often, they came from the student, and I had the flexibility to let them find their way along their own path. I had students choosing computer graphics problems outside of my research support, and I had a few who were well outside computer graphics. Conversely, I had a few students where I thought I defined good thesis topics for them, but the spark never ignited the necessary passion to carry through, and they left the program. In a few cases, I had to push them out, usually by cutting off funding. I did not find these situations pleasant at all and regarded them as my failures in judgment rather than the student's. In a few cases, I misjudged intelligence for passion. In others, I mistook strong programming prowess as an indicator of Ph.D. success. Students who might otherwise fall a bit short in academic and technical ability more often than not made it up with a focus on the goal, a willingness to put in the effort, and the determination to "rise above" their own background. I saw these students as reflections of my own experience, and I felt considerable pride for them when they succeeded in defending their work. Although I would like to recognize all these students personally, I think it is best not to name names. Nonetheless, many who did not stay on for a Ph.D. are quite successful without it. The Ph.Ds. got their ticket, but what they chose to do with it afterward has significant variability.

Many of my Ph.D. students didn't realize, at first, that they played a critical role in educating *me* about something new. As their advisor, I was not omniscient. Often, I saw a possible direction or could articulate what I didn't like about someone else's approach. From there, the student could do the expected background reading and *explain to me* what the constellation of problems and approaches looked like and where we might take it forward. Collaborative discussions, yes, but often a student-led idea that motivated achievement. In all, I have supervised or co-supervised 61 Ph.D. theses. The enormity of this is not the number per se, but rather the sheer quantity of topics, breadth, and scope that I have had an opportunity—no *requirement*—to learn beyond what I already knew. This is probably the essence of the term "Doctor of Philosophy" for me: the evolution of beliefs through persuasive arguments and the extension of intellect through the achievements of others. I will always be indebted to my students for greatly expanding my worldview. I do not think I would have ever invested the energy, had the ability, used the time, or sensed the need to absorb so much on my own.

Life Lessons

One of the most helpful lifelong guiding principles for me was to have at least one very long-term goal, easily articulated, and ripe with potentially interesting problems and approaches: for example, "How to make a computer *understand* human movement." However, this was no all-or-nothing "moonshot." It defined a goal, or even a dream, and not an absolute requirement. Along the way of trying to achieve that goal, my students and I learned a lot about ancillary but crucial parallel fields—sometimes outside computer science—such as anthropometry, anthropology, movement notations, dance theory, archaeology, cognitive science, psychology, human physiology, and human factors engineering. We never received funding to "just" do computer graphics; support was always to further multiple developments toward building interactive digital people, mostly for engineering uses. We learned to listen to what experts *and* users wanted to do but could not, found difficult, or were awkward, such as building physical cockpit and workplace mock-ups. Publishing our work in venues other than just the top computer graphics and animation conferences or journals was necessary to reach specialized communities where our work would matter.

I always engaged my Ph.D. students with the sponsors and projects I could obtain, even if ultimately the students did a tangential thesis and took other career directions. Many examples of this academic variability and flexibility exist. Tripp Becket, Diane Chi, John Granieri, Sonu Chopra, Rama Bindiganavale, and Liwei Zhao took positions at a local financial trading firm, Susquehanna International Group (SIG). Becket (Ph.D. 1996) used reinforcement learning for virtual human navigation simulation. His expertise jump-started the SIG machine-learning group on arbitrage processes: making profit from short-term variations in monetary conversion rates. Diane Chi (Ph.D. 1999), of EMOTE fame, joined his group at SIG. Granieri (Ph.D. 2000), who had been one of my core *Jack*

N. Badler, *On Raising a Digital Human*, Synthesis Lectures on Computer Science, https://doi.org/10.1007/978-3-031-63945-6_48

programmers after Phillips graduated in 1991, became a SIG real-time systems architect. Chopra (Ph.D. 1999), Bindiganavale (Ph.D. 2000), and Liwei Zhao (Ph.D. 2001)
joined programming teams at SIG. I once asked Granieri why SIG was keen on hiring
my Ph.D. students. His answer was crisp and telling: "We're looking for strong programmers who understand real-time finance applications. Computer animation students are
excellent programmers and work with real-time graphics and simulation. We teach them
the finance."

Every Ph.D. student deserves their own story. I have tried to include those with an
integral role in the evolution of our digital humans, but they are all important in their
achievements as well as my personal growth. I am pleased that 20% of my Ph.D. students
are women. This group includes the first woman Ph.D. in Penn's CIS department, Sakunthala Gnanamgari. Rebecca Mercuri abandoned audio simulation thesis ideas to pivot to
electronic voting. Mitch Marcus co-supervised, and since her 2001 Ph.D., she is now an
expert in that domain and contributes articles in the *Communications of the Association
for Computing Machinery*. Hyeongseok Ko produced our locomotion system for his Ph.D.
in 1994. He moved back to Seoul National University in South Korea and soon thereafter
founded a company specializing in computer systems for clothing and fashion design.
Min-Zhi Shao did his 1996 Ph.D. in graphics image synthesis, then joined Sony Pictures
Imageworks. One of his tasks was to clean up the body motion capture data taken from
Angelia Jolie for the animated movie *Beowulf*.

Charles Erignac (Ph.D. 2000) presented a relatively rare opportunity: a Ph.D. student
externally funded by the French government to do his research on whatever topic suited
him. He adapted the fluid aspects of maintenance procedures into a qualitative liquid simulation framework. After graduation, he joined Boeing and worked on controlling swarms
of drones. Other approaches to qualitative reasoning included Pearl Pu (Ph.D. 1989) on
causality in mechanisms and Paul Fishwick (Ph.D. 1986) on hierarchical process models.
Different questions about a process often required analyses at different levels of detail,
for example, sometimes at a physical level and other times just at a state or outcome level
[1]. Fishwick would eventually create and lead a DMD-like interdisciplinary program at
the University of Texas at Dallas.

Human reach, strength, and collision avoidance are interrelated. Several Ph.D.s investigated these topics, including James Korein (Ph.D. 1984) on geometric methods for arm
reach [2], Susanna Wei (Ph.D. 1990) on strength displays, Tarek Alameldin (Ph.D. 1991)
on computing reachable spaces [3], Wallace Ching (Ph.D. 1992) on arm motion planning
[4], Philip Lee (Ph.D. 1993 in Mechanical Engineering) on strength-guided motion [5],
Xinmin Zhao (Ph.D. 1996) on collision avoidance [6], and Deepak Tolani (Ph.D. 1998)
on analytic, non-iterative, inverse kinematics methods for arm reach [7]. Another thread
included semantic, context-influenced, or approximate methods for controlling the body
as a whole. Moon Jung (Ph.D. 1992) examined posture planning when human models
were asked to perform constrained reaches [8]. Libby Levison (Ph.D. 1996) contributed

the ideas of smart objects and object-specific reasoning, where objects "told" the human figure how they should be grasped for a designated purpose [9].

Ying Liu (Ph.D. 2003) had an unusual experience. One of our departmental Ph.D. requirements was to write and orally defend a survey paper on a current topic in computer science. As a graphics student, she chose to investigate human hair modeling. Unfortunately, Liu failed the oral exam. Although severely disappointed in herself, she reworked the topic, took the exam again, and passed. Subsequently, however, she pivoted her interests based on our Air Force work and completed her Ph.D. on GPU-accelerated human arm reach generation [10]. In 2005, she interviewed with Walt Disney Animation Studios. At the interview, they asked her to talk about her hair simulation survey. Liu received a job offer and was hired as—a hair animation specialist! She worked on the hair team for all the Disney animated features through 2022.

Many of my Ph.D. students parlayed their degrees into careers broader than or simply outside their topic specialization. Working with external sponsors, such as the Air Force, NASA, or the Army, gave my students more than just potential thesis topics. I felt that there was great value in having them make personal connections with people who had their own jobs to do but needed new tools or insights. Once again, I could appreciate the patience and kindness of my mentor, Henry Kramer. When I started with him, I knew neither assembly code nor computer vision. I was motivated to learn and discover.

Moreover, I didn't hide the *process* of obtaining funding from my students, as these personal connections were critical to both understanding potential sponsor needs and formulating compatible solutions. Students often accompanied me on scouting trips to potential sponsors, interim on-site reports, and final demonstrations. Unlike the more impersonal and anonymous reviews of National Science Foundation proposals, I wanted my students to appreciate that some known person thought their work was important enough to provide intellectual input as well as funding. I have had my share of NSF funding (and it was always welcomed!), but having real customers was exceptionally satisfying. Students could learn and exhibit personal poise, communicative effectiveness, and the self-satisfaction that comes from others appreciating their abilities and achievements. I think these attributes, not the thesis specifics, were the lasting imprint of doing a Ph.D. under my mentorship.

Along the way, I think I became more sensitive to individual needs and abilities, too. Sometimes these were implicit; other times they were more overt. Not every Ph.D. student I took on was able to complete the degree. Often, the main reason was that their career interests changed, and a Ph.D. was not needed nor worth the years of effort. Sometimes these were strong and definitely capable students, so personal choices mattered. Other cases were the opposite, where I could see that progress was slow or nonexistent, and we parted ways to avoid prolonging the agony. My overall Ph.D. student attrition rate was about 15%, or one in seven. Perhaps to offset this loss, I often adopted the "orphan" unsupported or disinterested Ph.D. students of other faculty. None of those cases turned out to be disappointing.

My faculty colleagues contributed much to my research, especially filling in areas where I felt less comfort and ability. Many of these collaborations resulted in co-supervised Ph.D. students. While recognizing them here, I want to give most of the credit to the co-supervisor. Ruzena Bajcsy's student, Chaim Broit (Ph.D. 1981), pursued 2D image registration algorithms, especially for medical applications where patient images need to be aligned to prior samples or generic anatomical atlases. Sakunthala Gnanamgari 's 1991 Ph.D. on automated information graph visualization had encouragement from Professor Howard Morgan of the Wharton School of Business at Penn. Dimitris Metaxas primarily supervised Suejung Huh's 2002 Ph.D. on cloth modeling and collision response. Liming Zhao (Ph.D. 2009) worked with novel motion graph concepts for human movement animation advocated by Professor Alla Safanova. Motion graphs smoothly transitioned one action or movement pattern into another. Michael Johns (Ph.D. 2007) worked with Penn Systems Professor Barry Silverman on human performance models and how they influence agent behaviors [11]. Rebecca Mercuri (Ph.D. 2001) studied safe and accurate voting systems with Mitch Marcus. Motivated by our work on military simulations, Professor Michael Greenwald co-supervised Jianping Shi (Ph.D. 2000). He attacked the complex topic of joining an ongoing large-scale simulation with local updates transmitted accurately to the user's client workstation [12]. Sonu Chopra's 1999 Ph.D. work on human perceptual attention modeling enjoyed Bonnie Webber's co-supervision.

I had other Ph.D. students who worked outside human modeling questions. Sometimes these were explorations of potential new research directions, and sometimes they arose from the student's own interests. One of my earliest Ph.D.s, Larry Grim (Ph.D. 1980), investigated reversible image compression algorithms because they were relevant to his work at DuPont. Dan Olsen (Ph.D. 1981) examined procedural development for interactive systems. Olsen went on to be a major contributor in the human–computer interaction community. Gerry Radack (Ph.D. 1994) looked into representations that would aid 2D "jigsaw" puzzle piece matching [13]. Our work with pottery sherds motivated this question. Tamar Granor (Ph.D. 1986), my Introduction to Programming course coordinator, implemented a thesis on user interface management systems (UIMS) when interactive apps were just emerging on the nascent Internet. Pearl Pu (Ph.D. 1989) looked at an early AI approach to causal reasoning for mechanisms. Isaac Rudomin (Ph.D. 1990) studied cloth simulation methods. Jugal Kalita (Ph.D. 1990), as a member of our ANIML seminar, investigated motion verb animation. Diana Dadamo (Ph.D. 1990) looked into presenting alternative motion exemplars to allow a user to choose apt motion controllers. Jianmin Zhao (Ph.D. 1993) reworked Joe O'Rourke's human positioning work to accommodate perspective and inverse kinematics algorithms more accurately. Eunyoung Koh (Ph.D. 1993) worked on decimating 3D mesh models to reduce their polygon count for displays. Paul Diefenbach (Ph.D. 1996) was one of the earliest proponents of pipeline processing with emergent GPUs for various graphics algorithms [14]. Francisco Azoula (Ph.D. 1996) investigated the causes and remedies for errors in transforming external body dimensional data into 3D digital human models. Pei-Hwa Ho (Ph.D. 1998) dove further into anthro-pometric scaling. Bond-Jay Ting (Ph.D. 1998) built curved surface body shapes based in B-splines.

Other novel situations arose and had to be treated with sensitivity. In at least one case, a student began to well up with tears as we were talking about her work. When I noticed, I asked her if she was having concerns about the topic. She admitted that she did, and we immediately pivoted to an alternative direction she chose. She did successfully complete her Ph.D. in a timely fashion. Another student requested that no military-sourced funds be used for his support. I usually had adequate funds from multiple sources, so I was able to satisfy his request. Since I received generous funding (especially in the 1990s, when I was graduating an average of two Ph.D.s a year), I could often adopt Ph.D. students who wanted to (or needed to) change advisors. Several women students were pregnant during their Ph.D. years. We rejoiced in our "lab" babies. Others had personal lifestyle changes. Sometimes I knew about them, sometimes not. Marriages among lab members happened.

A Ph.D. adds to a "family" tree related by "supervisor of" and "supervised by" links. John Mylopoulos was my Ph.D. advisor, and I knew well his 1970 Ph.D. supervisor, Theo Pavilidis at Princeton. Pavlidis was an early computer image processing pioneer. I used to joke, proudly, that "I came from a long line of Greeks." I had no curiosity about going back further until Mylopoulos' next Ph.D. after me, John Tsotsos, invited me to visit York University in Toronto, Canada. We wondered who supervised Pavlidis' 1964 Ph.D. at UC Berkeley. Easy enough to find out: that was Eliahu Ibrahim Jury, whose own Ph.D. came from Columbia in 1953. Jury's supervisor, John Ralph Ragazzini, received his Ph.D. from Columbia in 1941. Besides Jury, Ragazzini's students included other names recognizable in engineering circles: Rudolf Kalman of Kalman filter fame and Lofti Zadeh, known for fuzzy logic in computer science. I was surprised and pleased to learn that they are my academic "great-uncles." Tsotsos researched further back and found Ragazzini's supervisor, John B. Russell, Jr., an MIT Ph.D. in 1950. Apparently, Ragazzini was co-supervised by Lawrence Baker Arguimbau, who had a Ph.D. when Ragazzini earned his, making the dates work. Russell was a student of Vannevar Bush, a 1916 Ph.D. from MIT. Bush's Ph.D. supervisor was Arthur E. Kennelly. Kennelly does not appear to have a Ph.D. himself, though he was on the MIT faculty and was self-taught in Thomas Alva Edison's laboratory in West Orange, New Jersey. These relationships, while not biological, are strong contributors to understanding the process of conducting research. Although our topics and methods evolved with the times, my own "virtual DNA" heritage must have been an implicit factor in my own achievements. I owe them all my deepest appreciation.

I passed along to my own students what I could. Many went into university teaching. Eleven of these became academics in Ph.D.-granting institutions themselves, and they in turn graduated at least ninety "grandchildren" Ph.D.s of their own. So far, this set has a collective twenty-two "great-grandchildren" Ph.D.s. I am certain that this tree will continue to grow, flourish, and deepen. A Ph.D. can be a lifestyle choice.

As an educator as well as a researcher, I learned to motivate students (at all levels) by giving them interesting and challenging problems rather than just telling them what to do. Usually, better solutions and approaches were their own doing, empowering them and helping them confront novel situations later in their own careers. The one quote I kept on my desk for my entire Penn career was from General George Patton: "Never tell people how to do things. Tell them what to do, and they will surprise you with their ingenuity." I have had a generous share of surprises.

I enjoyed meeting my new DMD students at the start of their freshman year. Like most undergraduates, they were relieved to enter the university. But my favorite question to them at our first meeting was "What do you see yourself doing in four years?" Mostly this question took them aback; the quest to get into Penn had likely been stressful enough, and now I had the audacity to ask them what they wanted to do four years hence. The answers seemed to reflect peer pressure, perhaps unsurprisingly, rather than careful thought. As I went around the table, if one of the early responses was "I'd like to work at Pixar," I could be pretty certain I would hear that echoed by others. It sounds good, and would have been a reach goal for many. Another year everyone seemed to want to work in games. By senior year everyone had differentiated interests. They had to choose an individualized senior project which had some significant coding component. I enjoyed supervising this capstone course with Adam Mally. He brought current application, game, and programming expertise, while I brought depth and sometimes useful history. Some of the senior projects even fell outside computer graphics. Student interests change over four years. Whatever they chose, we tried to make it a positive, technically-relevant, personal achievement.

When I retired, I no longer gave lecture courses, but I did ask to continue supervising DMD senior projects alongside Adam Mally. My department Chair, Zack Ives, agreed. He also asked if I would like to take over supervising the Bachelor of Applied Science senior thesis course. These students have to write a twenty page minimum exposition on a topic of current computer science interest. I agreed to this role. Students choose their own topics and find a suitable and willing faculty advisor from anywhere in the university. I marshal them through the thesis process. I found this role rewarding for three reasons. First, I was able to codify requirements, expectations, schedule, and structure to guide thesis production. I did not want anyone throwing a hefty document at me that they hastily assembled the night before the due date. Second, I adopted a version of project submission I learned from Clark Erickson. By commenting on preliminary topic sentence outlines, allowing multiple revisions, and looking over at least two drafts, I could help students iterate and improve the document to a successful conclusion. Finally, the breadth of the theses covered all of computer science. Some of them were a challenge to read through, but I learned about many topics I wouldn't have even thought to consider. They were interesting to someone. They were also a weathervane for fads. One year I saw many blockchain theses. Another year had several on quantum computing. This year there were none of those, but a majority were engaged in AI and machine learning topics.

Once again, teaching with individualization has been my version of "lifelong learning." Launching undergraduates into the world is different than graduating Ph.D.s, but it is no less satisfying.

Although I have generally painted a rosy picture of my career, it was definitely not without failures and disappointments. Papers were rejected, proposals were not funded, and what I thought were good ideas didn't attract any followers. Even novel topics sometimes lay fallow for want of a student to take ownership. I have a lengthy list of project directions that went nowhere. I didn't keep records, but I guess that almost half of my proposals were rejected. Losing the first one hurt; after that I gradually came to accept occasional failure as part of the process. Viewed another way, I had to write two proposals for every one funded. The greatest disappointments were those I earnestly believed could make a difference. The funding world in the U.S. is complex and diverse. We were fortunate with *Jack* to tap into corporate support and the broad human factors community across multiple government branches.

I did have some notable failures. Martha Palmer and I, along with some other colleagues outside Penn, submitted an NSF proposal to further investigate creating human animations from verbs. Since one of the co-investigators was in a large corporate entity, we proposed to a very specific NSF program that supposedly facilitated technology transfer from universities to the private sector. The proposal received strong reviews, but could not be funded because we asked for more money than the program could support. Without the corporate partner, the proposal would not have fared as well in the regular NSF program, so we abandoned this route. Sometimes even a success can be disappointing. I joined a European Union proposal with my former Penn colleagues Bonnie Webber and Mark Steedman at Edinburgh. The grant was funded, but the EU regulations were incompatible with Penn's intellectual property rules, and so I had to withdraw. The most egregious failure, however, was my own fault. I had assembled a great team of colleagues at other universities and a major prime contractor, Lockheed-Martin, to respond to a new research program on Intelligent Tutoring Agents. The proposal took weeks to assemble with all the technical descriptions, organizations, and budgeting requirements. The final budget request was for US$40 million. I did some last-minute editing, got the final proposal through the Penn Research Office, and electronically submitted it just before 4 pm on the due date. Usually, the submission deadline is close of business, or 4 pm. Unfortunately, I had completely failed to notice that this program deadline was *noon*. We were automatically eliminated from consideration. There is no late option. After a month of depression (and therapy), I rationalized it away by saying the awardee was likely already "wired" into the proposal, and we wouldn't have been selected anyway. Nonetheless, it was the biggest professional mistake of my career.

My impetus to succeed did not arise from an innate wish to do so. Many extrinsic factors played a role, including opportunities, encounters with others, students who excelled and thus propelled me in new directions, and good fortune, too. In 2017, the College of Creative Studies at UCSB invited me back as their 50th Anniversary Commencement Speaker. I had heard many graduation speakers in my role as Penn's Engineering School Associate Dean, so I generally knew how these went with attempts at inspirational, uplifting, "follow your passion" topics. Most were otherwise rather forgettable.

Now, for the first time, I had to reflect on my own trajectory and distill some helpful nuggets. Two primary observations framed my speech. I first observed that "follow your passion" is the wrong message. One may choose a passion that produces a negative lifestyle, and sometimes one can't even figure out what one's passion might be. Passions are like fads: they come and go, are buffered by our surroundings and interactions, and change with age and maturity. My second observation was a response to the first. Playing off the College's "Creative Studies" theme, I explored how creativity might arise. In my own experience, creativity is sometimes motivated by desperation, by fear of the unknown, or by dire circumstances that demand novel responses. Fear drives adrenalin production, and literally gets our creative juices flowing. Some of my key life choices came from situations that appeared beyond my control; I had to try unexpected directions to escape. I don't think it was passion, but fear of failure, which drove me to take sideways leaps and find a new path.

My CCS mentor, Max Weiss, attended the graduation event. I hadn't seen him since the 1970s, and we both grew obviously older. Weiss was starting to look frail. Talking to him was the high point of my return visit to UCSB. I was able to thank him for the CCS positives that affected my career trajectory and remarked that even though he was right about my lack of ambition, it was because of my limited knowledge at that time. I could finally express my genuine appreciation for his role. We had a moment of mutual understanding and affection. Weiss passed away in 2022. I was grateful for the chance to meet with him that one last time in 2017.

By 2020, I had run out of funding for graduate students. My last Ph.D. student, Aline Normoyle, graduated in 2015. I was able to continue funding her as a postdoc researcher for three more years on an NSF grant on Sign Language evolution hosted through the Rochester Institute of Technology. Normoyle had computed various physical properties of human motions as part of her Ph.D. thesis. She then applied these measurements to videos of people performing sign language [15]. She also continued to work on my visualizing the past projects out of a welcome sense of purpose rather than for pay. Soon she joined Bryn Mawr College as an Assistant Professor of Computer Science and gained the freedom to pursue her own research directions.

Then came COVID-19. In March 2020, the first week of the pandemic lockdown, I had to have surgery. The bad news is that COVID-19 forced us to switch—immediately—to remote teaching via Zoom. However, the (sort of) good news is that I was in no shape to go into Penn and do stand-up teaching, anyway. During the 2020–2021 academic year, I taught from home, committing all my lectures to recorded Zoom video. This cemented my decision to retire from my tenured position at the end of June 2021. I desperately wanted to have a retirement "thank you" party to honor as many people as possible who passed through my academic life and enabled so many wonderful memories. Unfortunately, Penn pandemic era rules forbade indoor food events. It isn't a party at Penn without food!

We kept pushing off a possible date through 2022. Eventually, restrictions eased, and we could schedule the party for April 2023. The delay gave me time to formulate what I wanted to do. Hosting an event with hundreds of former students and colleagues would not be realistic. I recalled memorial symposia held in honor of colleagues who had passed away. How sad, I thought, that the honoree was not there to feel the warmth, respect, and attention themselves. I vowed that my retirement would be in lieu of a posthumous event, and let me express my appreciation while I was alive! This worked out beautifully. I invited representatives from most eras and levels and asked them to describe what effect passing through my lab had on *their* lives and careers. They did not disappoint. It was fifty years packed into a two-hour microcosm led by invited speakers Joseph O'Rourke, Cary Phillips, Diane Chi, Patrick Cozzi, Josh Nadel, Joseph Kider, and Funda Durupinar. Several other Ph.D., Master's, and DMD students attended, many of whom I hadn't seen in decades, including Tamar Granor, Steve Platt, Sonu Chopra, John Granieri, and Paul Diefenbach. It was the best celebration *and* reunion I could have imagined. In between opening remarks by Engineering Dean Vijay Kumar and closing commentary by CIS department Chair Zack Ives, I was emcee and got to introduce each of my guests. Maybe this is just evidence of me being a control freak, but being able to tell everyone—in person—my favorite anecdote or background for their invitation was completely exhilarating.

Among many memorable moments, one unexpectedly stood out. My son David had kindly offered to host a dinner at an excellent local restaurant the evening before the event for Ginny, me, and the invited speakers. He brought his family along: his wife Amy, and children Alison, Ben, and Will. We took seats along the banquet table, and it ended up that the youngest child, Will, who was 11, sat directly across from Josh Nadel. Josh was the youngest of the speakers, a major contributor to our Baures video, and a 3D game designer. Will talked games to Josh the whole evening, and Josh entertained him right back. The whole event left warm feelings all around.

These students and colleagues, their achievements and careers, and my memories helped me realize—several times—that my best long-term strategy was to let go of my work. Passions can blind one to making better choices. Launching *Jack* as a start-up was just one obvious case, but many times, we ceded incipient ideas and territory to others for their exploitation. MarchingOrder is a classic case. The DIS *Jack* that motivated DI-Guy and the MediSim trainer are also illustrative. That these academic explorations had consequences and launched careers beyond their origins is truly satisfying. Ph.D. students graduated and took their systems and achievements with them. We avoided the Pygmalion effect by not falling in love with our own creations. New ideas had room to flourish.

References

1. P. Fishwick and N. Badler. "Hierarchical reasoning: simulating complex processes over multiple levels of abstraction." First Annual Workshop on Robotics and Expert Systems, Instrument Society of America, Houston, TX, 1985, pp. 73-80.
2. J. Korein and N. Badler. "Techniques for goal-directed movement." IEEE Computer Graphics and Applications, Nov. 1982, pp. 71–81.
3. T. Alameldin, T. Sobh and N. Badler. "An adaptive and efficient system for computing the 3-D reachable workspace." IEEE International Conf. on Systems Engineering, Pittsburgh, PA, Aug. 1990, pp. 503–506.
4. W. Ching and N. Badler. "Fast motion planning for anthropometric figures with many degrees of freedom." Proceedings of IEEE International Conference on Robotics and Automation, May 1992, pp. 1052–1061.
5. P. Lee, N. Badler, S. Wei and J. Zhao. "Strength guided motion." Computer Graphics 24(4), 1990, pp. 253-262.
6. N. Badler, R. Bindiganavale, J. Granieri, S. Wei and X. Zhao. "Posture interpolation with collision avoidance." Proc. Computer Animation, Geneva, Switzerland, IEEE Computer Society Press, Los Alamitos, CA, 1994, pp. 13-20.
7. D. Tolani and N. Badler. "Real time human arm inverse kinematics." Presence 5(4), 1996, pp. 393-401.
8. M. Jung, N. Badler and T. Noma. "Collision-avoidance planning for task-level human motion based on a qualitative motion model." Proc. Pacific Graphics Conf., 1993.
9. B. Douville, L. Levison and N. Badler. "Task Level object grasping for simulated agents." Presence 5(4), 1996, pp. 416-430.
10. Y. Liu and N. Badler. "Real-time reach planning for animated characters using hardware acceleration." Computer Animation and Social Agents, IEEE Computer Society, New Brunswick, NJ, May 2003, pp. 86-93.
11. B. Silverman, M. Johns, J. Cornwell and K. O'Brien. "Human behavior models for agents in simulators and games: Part I: Enabling science with pmfserv." Presence: Teleoperators & Virtual Environments Vol. 15 (2), pp. 139–162.
12. J. Shi, N. Badler and M. Greenwald. "Joining a real-time simulation: parallel finite-state machines and hierarchical action level methods for mitigating lag time." Proc. 9th Conference on Computer Generated Forces, Orlando, FL, May, 2000.
13. G. Radack and N. Badler. "Jigsaw puzzle matching using a boundary-centered polar encoding." Computer Graphics and Image Processing, May 1982, pp. 1–17.
14. P. Diefenbach and N. Badler. "Multi-pass pipeline rendering: Realism for dynamic environments." Symposium on Interactive 3D Graphics, 1997.
15. A. Normoyle, B. Artacho, A. Savakis, A. Senghas, N. Badler, C. Occhino, S. Rothstein and M. Dye. "Open-source pipeline for skeletal modeling of Sign Language utterances from 2D video sources." 14th International Conference on Theoretical Issues in Sign Language Research (TISLR 14 Stage Presentation), 2022.